ARCHITECTS' PEOPLE

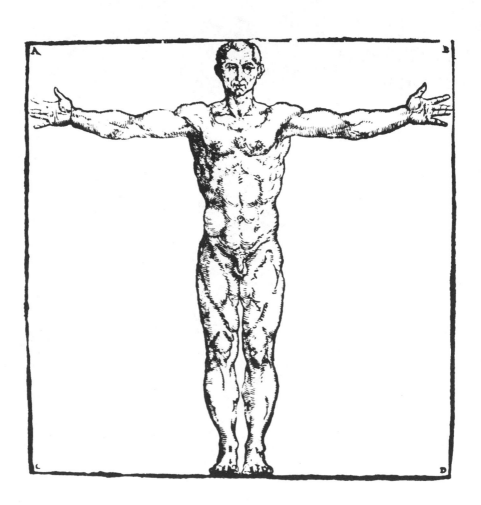

ARCHITECTS' PEOPLE

EDITED BY

Russell Ellis

AND

Dana Cuff

New York   Oxford
OXFORD UNIVERSITY PRESS
1989

Oxford University Press

Oxford   New York   Toronto
Delhi   Bombay   Calcutta   Madras   Karachi
Petaling Jaya   Singapore   Hong Kong   Tokyo
Nairobi   Dar es Salaam   Cape Town
Melbourne   Auckland

and associated companies in
Berlin   Ibadan

Published by Oxford University Press, Inc.,
200 Madison Avenue, New York, New York 10016

Oxford is a registered trademark of Oxford University Press

Architects' people / edited by Russell Ellis and Dana Cuff.
    p.   cm.
Bibliography: p.
Includes index.
ISBN 0-19-505495-4
1. Architecture—Human factors.   2. Architecture and society.
I. Ellis, Jr., William Russell.   II. Cuff, Dana, 1953–
NA2542.4.A68   1989
720'.1'03—dc19   88-25409   CIP

Printing (last digit): 9 8 7 6 5 4 3 2 1

Printed in the United States of America
on acid-free paper

Acknowledgments

Zoë, David, and Judith made the foundation. My mother, Marjorie Ellis, took the architect on daily trips through novel buildings. David Vincent and Lynn watched the sketching. Troy Duster and the gang at the Institute for the Study of Social Change had the best design ideas in Berkeley. Thelton Henderson's studio was crucial. Joyce Berry and Paul Schlotthauer at Oxford University Press were patient in inspecting the working drawings. Sol and Katie Fairston kept note that the work site was disorganized and the project over budget and overdue. Richard Ingersoll, Sylvia Russell, D'vora Treisman, and Deborah Stein handled change orders with care. "The Men" did the framing. My father, Russell, Sr., would have been pleased with the form.

R. E.

This editor's people, like those of the architect, are both actual and illusory, but influential nonetheless. Some very real readers gave me advice and encouragement throughout the project. Gerda Norvig has been my partner in the reader-writer duo through her sympathetic criticism, wisdom about the English language, and steady moral support. Diane Favro, Caroline Hinkley, and Beth Robertson maintained what seemed like tireless insight; by speaking to these women, my own voice has grown clearer. There are also guides whose work I admire, who provide the models by which I judge my own efforts in this and other projects. In this regard, Bob Gutman has been an unfailing companion. Russ Ellis and Roger Montgomery, as with other projects they've reviewed, have provoked good, clear thinking as this volume evolved. Lodged deeply in this editor's soul are three people whose lives interlace with mine in the complicated patterns that the everyday inspires. Kevin Daly, the architect I admire most, has been at my side throughout this project, teaching me new ways to look at the profession and celebrating every incremental achievement as if it were the final step. And present with me always are Ruth and Charlie Cuff, whose unwavering faith in my ability opened the world to me, and me to the world.

D. C.

Contents

III. The Subject's Identity

IV. Misfunctions and Revisions

Foreword

For an architectural historian beginning to practice when I did, twenty-five years ago, there were two classes of names that mattered professionally: architects, past and present, who designed buildings, and we historians who explicated these designing people and their work. Members of the two classes formed fateful liaisons in the process. We spoke of Michelangelo and James Ackerman in the same breath, and so too of other famous pairs of subject-interpreter like John Nash and John Summerson, Wright and Hitchcock, Venturi and Scully.

The only other names to be heard in our discourse were titled or independently consequential people who had caused the monuments of the grand canon of architectural history to be made—people, in the words of one of our own, Erwin Panofky, "whose prestige and initiative summons other men's work into being." The list included emperors, abbots, rich merchants, captains of industry, and the occasional heiress, philanthropist or poet. Justinian had been responsible for Hagia Sophia; Odo, Hugh of Semur, and Peter the Venerable for Cluny; and several hundred monied or celebrated folk for their own residences, distinguished enough in one respect or another to end up on the pages of history. There were the Palazzo Medici in Florence and the House of Jacques Coeur at Bourges, the Villa Cornaro and Castle Howard, and all those houses of the last two centuries that seemed to propel the livelier courses of modern architecture, the houses of the Shermans and the Robies, the Steiners and the Stoclets, the Tasels, Guells and Maireas.

Relationships between patrons and architects were not always pellucid. It seemed, at times, that the architect receded so far in the equation as to be dispensable, at least as far as history was concerned. It was Abbot Suger who took all the credit for inventing Gothic at St. Denis in the 1140s; if he ever consulted an architect, he did not think it necessary to mention the fact in that effusive account of his cleverness he penned so diligently. It was Herod's Temple, and Charlemagne's Chapel at Aachen, and Mark Twain's house at Hartford.

At other times the two actors were matched. The patron knew enough about architecture to sit down at the drafting board, so to speak, as an equal. This was supposed to be a singling fact about the Renaissance, and much could be written about the relative importance of the duke of Montefeltro versus the architects he employed, Laurana and Francesco di Giorgio, in the design of the great palace at Urbino. So too with the collaboration of Nicholas V and Alberti, Pius II and Rossellino, Sixtus V and Domenico Fontana.

The architect had ways of getting even. He could push aside the patron most effectively by working for himself, as Adam did at Adelphi Terrace and Nash on Regent Street. The neatest way to rid yourself of all partners in the conception of your building was to do a house for yourself. Sir John Soane's at Lincoln's Inn Fields in London was famous, and so in our day was Philip Johnson's at New Canaan. Failing such architectural self-love, you could contrive to disparage your patrons' wisdom and taste, or ignore the complaints your designs elicited from them. Wright put-downs of patrons were legion, and it was perfectly de rigueur in our circle back then to expect a patron who had scruples about a building he had just paid for to set them aside and learn to live with it. The prerequisite for this enormity was that the architect be famous, and the building be a work of art. The rest was shadowy: clumps of figures who only obscured the elemental agon of patron and architect, and would have made the historian's task untidy, not to say unglamorous, were he or she in the least inclined to pay serious attention to such specters.

What was their sort? Well, on the side of the patron, one was diffidently aware that the term could not be thought synonymous with "client." Andrew Carnegie was the patron of all those libraries, but their clients were the cities that accepted his largesse and the conditions attached to it. And that was another thing, this business of cities commissioning buildings. The story of the arch-protagonists in the act of design became mushy when the force that summoned the architect's work into being was some unindividuated corporate agency. Ernst May's great housing estates were done for the municipality of Frankfurt, but who could dramatize Frankfurt as his mate in creation in the same way that one might write of Neutra, say, collaborating with Dr. Lovell? The trick was to personalize the corporate client. Athens may have been the client of the Parthenon, but it was more engaging, if largely inaccurate, to put a face on the abstraction of the *polis* and pit Pericles against Phidias—the statesman against the director of the works.

On the side of the architect there were ambiguities too, if you let them insinuate themselves in the grand epic of monument-making. The incidental stagy photograph of a beaux-arts master in his atelier, for instance, could be a faint reminder that behind every pedigreed design stood lesser pens who traced what they did not conceive, and may have done much more for which the star system would never allow them any credit. Should the apprentice himself attain that rarified realm of stars, the effort to isolate his individual touch in the days of training might be justified. What did Wright bring to Sullivan's projects during his brief apprenticeship in the older man's office? What did the young Sullivan do for Frank Furness some twenty years earlier? That was newsworthy. Elsewhere, sorting out who contributed what in the process of developing the design and staging its formal presentation could not be countenanced—not only because the attempt was bound to prove, for the historian, immensely laborious and ultimately inconclusive, but also because the collaborative drive behind the architect's official schemes had little professional appeal.

One might pretend otherwise of course. Collaboration was, after all, a buzzword of modernism—the Gropius wing at any rate. It was the C in TAC, though one would be hard put today to remember any name but Gropius' in that propagandistic enterprise. No, far more comfortable than the idea of a team of equals fretting over a design solution was the idea of the architect as the guiding spirit of the *Gesamtkunstwerk*. Painters, sculptors, craftsmen of various sorts, a constellation of talents fixed and empowered by the architect's vision—*that* was an acceptable extension of creative company, as long as the primacy of the architect did not suffer undue compromise.

Now the *makers* of the building—carpenters and masons and roofers and plasterers—crowded around the architect as well. How close was a matter of the architect's choosing, or of his moment in history.

In our own sphere it was common knowledge that there were architects who cared how their buildings were made and those who did not. This reputation had a certain cachet—and also a theoretical spin-off. Gaudi, we knew, had been almost mystically involved in the making of his buildings, Mies had fretted about those beautifully burnished corners of his metal cages, and Louis Kahn had transformed concrete detailing into a modern metaphor of the classical orders. Of other celebrated contemporaries, it was damning enough to say, "The building is already falling apart."

The loftier debate, however, concerned the ancient conflict between the architect as craftsman and the architect as conceptual artist. Both popular and scholarly consensus had it that the medieval architect was the paradigm for the first, rising as he did from within the guilds, knowing how to carve a capital and spin a bit of tracery before he was allowed to design. Setting aside the hyperbole, all that fevered Ruskinian rhapsodizing about the carver of ornament as the true edifying spirit of cathedrals, enough hard evidence in fact existed to prove that no fine distinctions were made in the Middle Ages between builders, or even mere masons, and architects. The Gothic architect sought recognition as much through his mechanical prowess as through his design, and he was expected to take active interest in the building process, even to the point of supplying the building materials.

The Renaissance architect would have none of this. Again the actual state of affairs tends to be inflated—Brunelleschi, after all, started out as a goldsmith, and Palladio as a stonemason—but that something essential for the practice of architecture had been established in northern Italy during the fifteenth century could not be in doubt. Alberti had said it first. In the opening paragraphs of his treatise on architecture he had canonized the purity of the designing mind, and had set it above the toil of building.

> But before I proceed further, it will not be improper to explain what he is that I allow to be an architect: for it is not a carpenter or joiner that I rank with the greatest masters in other sciences; the manual operator being no more than an instrument to the architect. Him I call an architect who by sure and wonderful art and method is able both with thought and invention to devise, and with execution to complete, all those works which . . . can with the greatest beauty, be adapted to the uses of mankind.

That fateful disjunction persevered to the present. It was enshrined in the code of ethics of the American Institute of Architects, which prohibited its members for a whole century, until the about-face of the late seventies, from being engaged in building.

For all that, no architect could be divorced altogether from the human factors of his designs, and the fringes of the history of architecture teem with those unsung casts of thousands who cut the voussoirs for arches and make the formwork for concrete, and toil on their scaffold perches until rain spouts are sealed and secured and the last roof tile is in place. This is the story of the guild and the lodge, and "the secret of the masons." The workforce of a generational artifact like a cathedral outlived its architects. A special mystique surrounds the medieval *Bauhütte* or *chantier*; or the long uninterrupted line of the builders of St. Peter's in Rome, the confraternity of the *petrini*, whose ultimate allegiance is to its fabric.

But notwithstanding some showy reminders—signature scenes of the building crafts in the stained glass windows of Chartres, for instance—the workforce had no claims on the history of architecture. And neither did those who inhabited and used the canonical monuments. Adapted to the uses of mankind, Alberti had said. Well, yes, certainly. Buildings are meant to be used—that is what the colossal enterprise of patrons and clients, architects and builders, is all about. But I do not remember being taught that the architectural historian had to entertain any particular brief for the user, unless it be in an emblematic sort of way. In Hagia Sophia the galleries were for "the unbaptized," we would say; or we would mention the jousting tournaments in Bramante's Cortile del Belvedere, and flash on the screen Etienne Du Perac's engraving of 1565 which showed one such tournament in progress.

Users as the filler of architectural space figured intermittently in architects' talk as well, with memorable effect in rare instances. Bernini saw his piazza in front of St. Peter's as the embodiment of Mother Church who holds in those tremendous arms "Catholics in order to reinforce their beliefs, heretics to reunite them with [her], and agnostics to reveal to them the true faith"; and Charles Garnier's appreciation of the attending crowds as a spectacle in their own right became the explicit rationale for the "broad and monumental arrangements with vast and commodious stairways" at his Opera in Paris.

Modernist rhetoric waxed eloquent about the needs of users. It represented architecture as the vehicle of social welfare and set public housing as the highest priority of architecture. But there was no question of consulting with the user of housing estates during the course of their design. No one bothered to explain why, since the picture was too obvious. Users were not a stable or coherent entity. And users did not know what they wanted or, more importantly, what they *should* have. Their collective needs, interpreted by the architect and the sponsoring agency, would be codified in the "program"—as had been the case with hospitals, schools, and prisons in the past. The fit might not be comfortable at first. The setting might appear alien to our habitual ways. The fault was with our habits. We would learn to adjust to the new *Wohnkultur* because it was based on rationally derived standards. *Existenzminimum*, the space allocation that conditioned our living unit, was a scientific datum.

But then architecture never really tried to come to terms with users, even when the user was none other than the client—so why should the historian? The user, if you bothered to think about it, was in the way of the design except in the most abstract sense. As professor-architect Otto Friedrich Silenius put it in Evelyn Waugh's *Decline and Fall*: "The problem of architecture as I see it is the elimination of the human element from the consideration of form." In real life the Peter Eisenmans of today still express this same sentiment, as Dana Cuff brings out in her interviews for this book.

Now if such modernist declarations appear a trifle too frank to applaud without embarrassment, their faith is nonetheless buried just below the cultivated surface of earlier periods like the Renaissance. Was a Palladio villa a proper reflection of how its owners spent their days, or wanted to? Were they not, too, the innocent recipients of a *Wohnkultur?* The difference lay in the architect's phantasm—the *imagined* user designed to suit the architecture. For the early modernist, committed or forced as he was to deal with a mass-culture clientele, that was the heroic Worker, an updated postwar version of the nineteenth-century reformist's "new man." For the Renaissance architect, it was the noble-minded Humanist.

The fit between the phantasm and the architecture was devised accordingly. The Worker must have sunlight and air, windows and entrances where they made sense, and to hell with symmetry, no-frills walls and roofs, and to hell with taste, new materials, streamlined, compact ktichens, and bathrooms with up-to-the-minute fixtures. The Humanist must have the benefit of classical order for which the orders were a ready-made vehicle—and to hell with the comforts of the body or the rhythms of daily life.

So the modernist spun out his mythology: the gods Licht, Luft and Sonnenschein; the lands of Sotsgorod and Neue Sachlichkeit and Ville Radieuse; monsters called Wolkenbugel and Dom-ino and Zeilenbau. And there and so we must henceforth live. Gropius informed us of that in 1935:

> The intellectual groundwork of a new architecture is already established . . . the bench-tests of its components have now been completed. There remains the task of imbuing the community with a consciousness of it and its essential rightness. . . . The ethical necessity of the New Architecture can no longer be called in doubt.

So the Renaissance architect had spun out *his* mythology, and had made us live by its directives for several centuries.

The historian, always declaratively above the frey, could record and appreciate these comings and goings. None of this had anything to do with real human beings, of course; how they felt or what they wanted. Architecture was not that sort of game. Human needs, human nature, did not require fundamental changes in architecture—or that was not how you conceived of its history. Architectural revolutions required the redesign of humanity.

That is why, at Yale in the late fifties, we read with equal commitment Gropius' *The New Architecture and the Bauhaus* and Geoffrey Scott's immortal essay on the classical phantasm and its world, *The Architecture of*

Humanism. According to Gropus, it was enough to keep your mind centered on function: the design would take care of itself; and the occupant, sooner or later, would see the logic of the architect's way. According to Scott, we did not simply occupy architecture, we *were* architecture. It was our business to transcribe ourselves into its terms, to make masses, spaces and lines of architecture respond to our ideal stability and ideal movement. Stability was "humanized dynamics"; proportion, "a preference in bodily sensation." Ultimately architecture was nothing other than "a humanized pattern of the world, a scheme of forms on which our life reflects its clarified image." And no system of design was better capable of such clarification than classical architecture.

This business of architecture measured out to human limbs—architecture as the assembly of units based on the length of an arm or the height of a head—had an undying appeal; it seemed to make up for the willful way in which architects chose to ignore what humans actually did, or would have liked to do, in the designed environment. In Renaissance treatises there were images of a whole heraldic race of naked men awkwardly fitted into circles and squares, all of them stemming from Vitruvius' peculiar assertion that "if a man be placed flat on his back, with his hands and feet extended, and a pair of compasses centred at his navel, the fingers and toes of his two hands and feet will touch the circumference of a circle described therefrom"; and further, that the same spread-eagled man would also yield a perfect square, in as much as "the distance from the soles of the feet to the top of the head" would equal the breadth of the outstretched arms. Even some modernists, function or no function, succumbed to these human measuring rods. Le Corbusier's modulor, reaching heroically upward with one ramrod arm, is certainly as familiar as the architect's own likeness in photographs.

Naked men to one side, what all this architectural "humanism" came down to in *practical* terms was a system of proportions in which all parts of a building derived from multiples or fractions of a key unit. In Greek temples this unit was the diameter of the free-standing column; in Chinese architecture, a Sung manual of about 1100 tells us, it was the end elevation of the horizontal corbel bracket arm or the *hua kung*. Le Corbusier's gesturing hero gives us a series of predetermined lengths generated by the application of the golden section (0.618) to the height of a six-foot man.

To be sure, the presumption is always there that these proportional systems are compatible with the human frame moving through buildings: the classical intercolumniation would be wide enough for a person to walk through unencumbered, and so on. The attempt is even made to coordinate architectural dimensions with specific human activities. In a passage of the *Mu Ching* (Timberwork Manual), the work of Yu Hao, a famous master builder of the Sung period, three types of ramps are distinguished: steep, easy-going, and intermediate. Their gradients, in palaces, are based on a unit derived from imperial litters.

> Steep ramps are ramps for ascending where the leading and trailing bearers have to extend their arms fully down and up respectively. Easy-going ramps are those for which the leaders use elbow length and the trailers shoulder height; intermediate ones are negotiated by the leaders with down-stretched arms and trailers at shoulder height.

A person, a litter bearer, Vitruvius' "well-shaped man." Ideal human types, ideally translated into columns and ramps. In reality, children use buildings; the elderly, the disabled use them; short and tall people use them; workers and employers use them. So the shape of users varies, and so do their physical performance and their purposes. How could such specificity be reduced to an architectural system of unarguable proportional relationships? Have we nothing to say about any of this? Could we wrestle down the phantasm and install ourselves in this place?

To speak for this architectural historian, the revelation that opened up such questions was a little book on Le Corbusier's pace-setting twenties settlement of low-cost housing at Pessac, near Bordeaux. The book, by Philippe Boudon, came out in 1969, and recorded in simple words and photographs an astounding rebellion. The architect had provided people with stark, modernist, value-free containers, which were at once technologically justified "machines to live in" and esthetically disposed, cubist arrangements of closed and open volumes. Here working class families were to install themselves, and fit in the quotidian details of a modern life.

What they did instead was to rebuild the containers, little by little, so that the architecture would accommodate their actual needs. They narrowed and framed the windows; they blocked off the empty spaces beneath the trademark Corbusian stilts that held aloft the house cubes; they sealed the see-through roof terraces; they appropriated their unit with their favorite trees and shrubs; and they built common sheds, unmindful that these would impair the purity of the design. The users had struck back at the most celebrated architect and planner of modern times. Had not he himself once said resignedly, in speaking of Pessac, "You know, it is always life that is right and the architect who is wrong'?

By the time Boudon's *Lived-in Architecture*, this extraordinary treatise on the user as designer, had appeared, I had left Yale for Berkeley, and had traded a department of the history of art for a school of architecture. It was a time of radical revaluation, a time of angst and social violence, and my new professinal home resonated with its intensity. Buildings were now protagonists in public civil struggles. Sproul Hall was *occupied*. Would-be users fiercely fought the National Guard to abort the client's plans for the liberated "People's Park," a piece of university real estate recently cleared of some old houses for the construction of high-rise dormitories. At the school there was talk of "participatory design"; open studios were set up to bring the benefits of architecture directly to the unempowered, to work with Indian reservations and poor Black communities in East Oakland on the upgrading of their environment. It was a new article of professional faith that patients had as much right to determine what sort of hospitals were built as did administrators, and that the disabled and the elderly had special requirements which must be heeded in the design of any public amenity. I remember observing with amazement the installation of crude ramps and the hollowing of sidewalk crubs, to make the city of Berkeley accessible to wheelchair traffic.

Specialized studies of the social and behavioral sciences now sought to quantify user preferences and evaluate the performance of environmental design, as Roger Montgomery describes in the last paper of this book. The ideal Vitruvian man and the Modulor retreated in favor of people with actual dimensions. The first edition of John Croney's *Anthropometry for Designers* was published in 1971. He wrote:

> It is impossible to correlate artistic and industrial efforts to help manufacture things of use without gathering information about man first of all. If the information is to be worthwhile and apposite one of the things man must do is to measure himself. . . . The most successful environmental system would be a replica of his own system. [To be a successful designer,] you must be conversant with different limitations of human performance.

In the most radical casting of these challenges, there were those who concluded that architecture was too important an intervention to be left to architects. Friedrich Hundertwasser's "Manifesto for the Boycotting of Architecture" is an authentic document of its time.

> Every man has the right to build the way he wants. . . . Everybody would be entitled to build his own four walls and be responsible for them. Present-day architecture is criminally sterile. The reason is that building stops when the client enters his residence, yet that is precisely when it should begin, and grow like skin on a human organism.

The effect of all this on my own thinking as a historian of architecture was tonic. My research and published product may have been slow to respond—it is hard to shake loose the hallowed practices of the academy—but in my own reading and teaching I crossed the threshold to a broad, adventurous realm that had been kept away from me. It was as if my field waited to be reinvented.

What did happen in those Renaissance palaces of the Medici and the Farnese, I now wanted to know; who slept where and what rituals and family ranks appropriated the rooms of the piano nobile? I recognized that history had always interpreted the sovereign house from the viewpoint of the lord, not the staff and the servants; the church, from the viewpoint of the learned celebrant rather than the common worshipper. Who built Greenwich Hospital; where did the stone come from and who was responsible for bringing it? What happened to Hagia Sophia after the fall of Constantinople; what did its Ottoman heir make of that signifying system of holy spaces and the gold-backed mosaics? Who bought the Adelphi houses; who were the middlemen and financiers? Ignorant as yet of the stirring Foucault inquiries, of *Madness and Civilization* or *Discipline and Punish*, I wandered unaided into the circles of compulsive occupancy. I set out, before class lectures on corresponding architectural topics, to find out what I could about being an inmate of the lunatic asylum in Bayazid II's great *kulliye* at Edirne or a prisoner in New York's Tombs.

It was only a matter of time before some of us across the country were fully converted to Braudelian dogma, the writing of history from the bottom up. It seemed inconceivable now that one would not embrace the view that what is fascinating in history is "the extent to which it can explain the life of men as it is being woven before our eyes, with its acquiescences and reticences, its refusals, complicities, or surrenders when confronted with change or tradition." And what better field than architecture for these human revelations since everything we ever do is done in a built or delimited frame—a house or a market, a farm or a skyscraper, a birth ward or a cemetery?

This charge is indeed overpowering. But it cannot, so it seems to me, be forever avoided. The sixties are over, we know. The revolution has been neutered. It is rare to hear talk of users now, even in Berkeley. Many architects are content yet again to design for peers and for the elegant phantasms of their gift. Still, we on the outside must persist. We must remind them that architecture floods the reaches of its creation, that architectural discourse refuses to be trapped within the gilded, hermetic symposia of design priests and their verifiers.

That is why *Architects' People* is an important book. It insists that the business of architecture is that of "person-environment relations." It seeks to reexamine that unresolved, unresolvable, dilemma of the profession: where does the soul of the architect seek its peace between the sublimity of the "utopian designer" and the more basic satisfaction of the "social servant"?

The editors, not themselves architects, are wise enough to recognize the splendid and inherent arbitrariness of the act of design. They strike precisely at the architect's existential core, where Formality and Use must fight it out. There is a mystery in the designer's course, beyond programs and budgets and users, a mystery which cannot be denied or deflected by those on the outside. The architected form has the power to outlast whatever restrictions may have conditioned its genesis—to become an object of universal admiration free of time, place, and circumstance.

But in the final reckoning, the fate of this talisman is in our hands. We are the ones who turn churches into barns, textile mills into convention centers. We determine what to pull down and what to cherish as the expression of our national pride. We are the ghosts in empty rooms of southern mansions, the rude, consuming crowds on the exquisite tracks of the Taj Mahal. We are architects' people. We cannot, would not, do without them. Without us, they have no cause to celebrate their ancient art.

Spiro Kostof

Contributors

Giandomenico Amendola, School of Architecture, University of Bari

Kent Bloomer, School of Architecture, Yale University

Dana Cuff, School of Urban and Regional Planning and School of Architecture, University of Southern California

Johanna Drucker, Mellon Faculty Fellow, Harvard University; Program in Arts and Performance, University of Texas at Dallas.

Russell Ellis, Department of Architecture, Institute for the Study of Social Change, University of California, Berkeley

Diane Favro, Graduate School of Architecture and Urban Planning, University of California, Los Angeles

Paul Groth, Department of Architecture and Landscape Architecture, University of California, Berkeley

Robert Gutman, School of Architecture, Princeton University and Department of Sociology, Rutgers University

Joseph Juhasz, College of Design and Planning, University of Colorado at Boulder

Lars Lerup, Department of Architecture, University of California, Berkeley

Raymond Lifchez, Department of Architecture, University of California, Berkeley

Roger Montgomery, Dean, College of Environmental Design, University of California, Berkeley

ARCHITECTS'
PEOPLE

Introduction

RUSSELL ELLIS and DANA CUFF

The purpose of this book is to bring into one place a concentrated discussion of people and architecture: reflections on the people who are imagined to occupy the buildings conceived, designed, and executed by architects. Inevitably, one of our topics is the thinking of architects as reflected in their speaking, writing, and designing. But more basically, the goal of this collection is an understanding of the architect's imaginings as they imply the social worlds of buildings we do or could inhabit.

Architects are the professional keepers of the knowledge and skills that render the built environment. Theirs is the historic charge to conceive and reconceive design. At the level of imagination and utility they accept this charge and undergo extensive formal training that prepares them to exercise it. Presumably, they learn somewhere how people do or want to live. If particular architects are uninterested in what people do or want, at the very least they must harbor some notions of how people might live—perhaps as static admirers of buildings as art. This book proceeds on the assumption that it is impossible to design a building without some conception of human activity in and around it, and it investigates those conceptions.

Why does our topic matter? An old Italian immigrant to the United States was asked by a reporter how he had the prescience to buy up major portions of California's Monterey peninsula many years ago at very cheap prices. His response was something like, "Well, I kinda figured there was going to be a lot more people and no more land." Ignoring the interstellar prospects, he was, of course, right. For us, as a species on the finite site earth, the implication of this fact is clear. Increasingly, we will be faced with the necessity of communicating to ourselves how we want to conduct our environmental lives, in what settings, in what ways, and in what array. A permanent condition of increasingly scarce spatial resources will require particularly acute interpreters of our emergent spatial practices and desires. Socially imaginative leadership must become a more prominent feature of designers' work.

Along with the realization that we perpetually face "a lot more people and no more land" must come a concern for how we should build on and manage the land. Many new situations that face architects today demonstrate the need for creative, socially alert responses. To take just two examples, consider the environmental activism of certain special populations on the American scene and the increased demand from distant cultures for architectural services.

The term *special population* has been used among social scientists and designers alike, seemingly to refer to any group that is not part of the imagined "normal" population. Children, the elderly, ethnic communities, gays, the disabled, and nonnuclear households are a few such special populations. As Dolores Hayden argues, the American dream and the norm of a nuclear family, which still drive housing production, leave out more than half of America's population.[1] Who are these people? As the elderly, the disabled, and most recently, the homeless have shown, they can organize to make certain they become part of the architects' people. The first two groups have been particularly effective in setting policies and programs so that their environmental needs are not left to imagination alone.

In the second case, architects increasingly find themselves working in foreign countries and in cultures unlike their own. This point was brought home recently when some students asked for information that might help with their design project: a multifamily housing complex in Saudi Arabia. In particular, they wondered about the role of women in that culture and what it meant for daily living patterns. Were women trying to overcome their traditional roles? These bright, conscientious students imagined the Saudi women as if they themselves had been transplanted into the Saudi culture. But their imaginal resources were limited. If we are not to create more Brasília's (recommended by Brazil's own dissatisfaction and desire to create another capital) we must be more conscious of context—not only of the heritage of physical artifacts, but of the life inscribed there. If our mental actors are ethnocentric, we must learn to develop new ones.

Until the present, there have been few reasons to examine the architects' people.[2] Tribal, traditional, monarchic, and autocratic societies typically provide clear rituals and rules out of which the built environment is extruded. For centuries, power, policy, and habit in legal–rational societies (particular Western democracies) have provided fairly stable instructions to the makers of the built world: instructions that produced some semblance of isomorphy between class, culture, and housed lives. In recent decades, although environmental facts may not have changed radically, the broad social impulse to change the conditions of architectural conception and execution has.

Robert Sommer observes that our discussion in the last three decades of "user needs" has precedents in the consumer movement.[3] Despite the reef-florescence of participatory democracy and citizen participation, social scientists have dominated the discussion about the features of people that ought to be or might be attended by architects. Few have asked architects what has guided their own thinking on this topic.[4] If our premise is correct that no building can be designed without some noisy or quiet conception of its prospective use, then we can discover some set of architects' conceptions of people and social life attendant to their designing. This book seeks to display some of those conceptions, to sharpen the discussion, and to see where it may lead.

Since social scientists started becoming architectural specialists, architects have had to defend against philistine intrusions into pure design. It has been clear—especially postmodernistically clear—that human action considerations have all the appeal of a draft horse in the Preakness. "The program" is unavoidable, even for the most abstrusely aesthetic architect. Although the topic at hand is not, strictly speaking, programming, "the art (and craft) defense"[5] against a building's function is a means to put the program in its place.

> Good architects have always had their sources, no doubt, but instead of owning up to them, they have often talked pseudoscience about the building as the inevitable expression of its program. Programs, sites, and budgets don't design buildings, though. Architects do, and as anyone who has ever tried it knows in his or her heart, you can't derive architectural form from any amount of analyses of context or program or anything else. There has to be an arbitrary formal gesture at some point or the design process can never begin.[6]

This position is convincing and may indeed be the true challenge of good design. But our topic stands. Buildings are for people, and it is never irrelevant to ask after the relevant thoughts and actions of architects on the "peopled" aspects of their designs. Indeed, it is increasingly imperative.

The Hidden Actor in Architecture

> If it is true that each of us carries around an implicit theory of human personality or behavior based on continuing experiences, and which we use to gauge and evaluate people we first meet, it is probably no less true that architects, designers, and planners have built into this theory something about people in relation to places and spaces.
>
> HAROLD PROSHANSKY

A few years ago, following a lecture provocatively entitled "The Ectoplasm of Buildings," one of us managed to catch the interest of an architect in the audience who felt he understood what the lecture was getting at. Over the course of a long conversation he volunteered the following:

> Oh yes, I've dreamed I was a building, and I changed myself in the dream. I've also dreamed I was an element in the landscape.
>
> When I'm "looking around" in a two-story space, my consciousness is everywhere. When it's a one-story space, I'm looking from about eye level.
>
> When I'm doing urban type work, I tend to draw Gordon Cullen people. I hate Jacoby people.

Imagine! Through the bubbling activity of dream life actually becoming building, parts of sites, and changing oneself! Prospective buildings were animated through him and he experienced their structures by becoming them. One might wonder on behalf of whom his self was used in this experiencing and in response to what the self-building was changed. More talk with him provided answers. The dreaming was a design activity in an architectonic sense. Anthropomorphizing the building through himself was a way of establishing a dialogue between the emergent structure and his own evolving ideas as a designer.

It became clearer how some architects can seriously say things like "the building wants to be that tall."[7] If you are the building it can have volition. If your consciousness permeates a two-story space and, as designed, it doesn't make sense, it can want to be a different shape or volume, and push itself out or contract until it is comfortable. The clue here is that the architect, awake, describes a dreaming eye floating disembodied and assembling information for the designer and the designed. In the case of this architect, when he does "people" his designs, he prefers a particular style of architecturally rendered people who will inhabit the designed results.

Working with architectural designers and students, we have noticed a seeing-linked ability—"imaginative self-projection"[8] that seems to focus architects' visual awareness of the experience of a proposed design. An observation of Robert Maxwell's is especially suggestive:

> We look at plans and we imagine doors swinging, drawers being pulled, corridors full of racing feet, people falling down unexpected steps or jamming on landings. We compare the drawing with similar drawings or with actual buildings in our memory store and we say, with considerable confidence, it wouldn't work, the circulations would cross, the corridor is too narrow, the room is claustrophic, and WC's are too far away, the waiting space is intimidating, and so on.
>
> In other words we attribute to the drawings or models operational qualities based on our own experience, and assess the performance which we would expect, imagining the building built and us in it.[9]

Over and over again in design sessions or studio critiques, schemes are referred to as if their spaces were occupied.[10] The traditional architectural program encourages this tendency through articulation of the prospective design's occupants and their activities: a family of five, low-cost housing, a day care center, or a factory for the production of electronic equipment. For some experienced architects such a program will prod their visual imagination beyond the mere square footage or equipment specified. For others, the life to be housed is indelibly etched into an architectural portrait, as Lerup and Bloomer discuss in their chapters. In greater measure, however, the occupants that architects project into their designs and plans are often empty forms, or nonpeople, as Groth puts it. Design education itself has only recently begun to address in any immediate detail the *content* of human activity.[11] Ray Lifchez's integration of fictional characters and studio teaching is a model alternative.

One can easily understand the necessary abstractness of actors and action in the frequent instances where the inhabitants of proposed buildings are anonymous (e.g., in an organizational bureaucracy). We do not mean to suggest that these actual inhabitants should be ignored. Even when we can locate them or their representatives, what we learn will be tempered by our mental dwellers. Gutman's study of Louis Kahn demonstrates how active the architect's own progeny can be. The authors in this volume explore the features of the implicit actor who lurks in the designer's imagination. What sort of character is this model actor who falls down steps, for whom drawers jam, who experiences claustrophobia? Alfred Schutz's critical formulation, applied to social science theory, is remarkably applicable to these questions.

The Architect's Homunculus

Schutz contended that the social scientists' peculiar failing is that they provide no "here," no living locus, to ground the meaningful social life they attempt to understand. Even the participant observer in the field setting, he argued, only temporarily drops the scientific attitude to make close experiential contact with the group being studied. Presumably, this temporary adoption of a position in the field gives the participant some understanding of the observed behavior's meaning to the actors. But ultimately, the scientific attitude requires not only understanding, but the actual "construction of some model of the social world and the actors in it." To make sense of observed behavior, the scientist invents an actor with features related to the phenomenon under investigation (e.g., status occupancy, family life, power, etc.). The invented actor is supplied with a "fictitious consciousness." Schutz imagines this invented actor as a kind of puppet, or *homunculus*, and in a critique of the social sciences he contends:

> these models of actors are not human beings living within their biographical situation in the social world of everyday life. Strictly speaking, they do not have any biography or any history, and the situation into which they are placed is not a situation defined by them but by their creator, the social scientist. He has created these puppets or homunculi to manipulate them for his purpose. A merely specious conscious is imputed to them . . . . which is constructed in such a way that its presupposed stock of knowledge at hand . . . would make actions originating from it subjectively understandable, provided that these actions were performed by real actors within the social world. But the puppet and his artificial consciousness is not subjected to the ontological condition of human beings. The homunculus was not born, he does not grow up, and he will not die; he has no hopes and no fears; he does not know anxiety as the chief motive of all his deeds. He is not free in the sense that his acting could transgress the limits his creator, the social scientist, has predetermined. He cannot, therefore, have other conflicts of interests and motives than those the social scientist has imputed to him. He cannot err, if to err is not his typical destiny. He cannot choose, except among the alternatives the social scientist has put before him. . . .[12]

What is interesting about Schutz's observations is his examination of a way of thinking by which students of society model real phenomena in order to understand them. We propose that architects, because they focus on buildings, have their own distinctive tendencies as they model their inhabitants. It has been our informal observation over the years that architects often assume that meaning and potential action reside in things. (This observation is corroborated in some of the interviews with contemporary architects included in this volume.) They do not usually make a clear connection between human interchange and the resultant deposit of meaning in "mere" things. The architect's homunculus is sometimes featureless, emerging first as disembodied actions among a design's details. Actions float free. As in Maxwell's comment, there are "doors swinging, drawers being pulled, corridors full of racing feet," and circulations that cross. Unlike the social scientist's invented actor, consciousness does not appear to be a common feature of the architect's homunculi. Indeed, in moving through designs with some architects one gets the impression that an indistinctly motivated lump of somatic stuff—born in and taking shape in bubble diagrams—is being conducted via arrows along paths of circulation to loci of living, eating, and bathing. This little puppet, though animated by the designer, tends to be passive and unobtrusive of the design's flow.

It is not extreme to say that for many designers, architecture *creates* the homunculus. The emergent plan generates the action of the somatic lump, and as the design takes on added features, so does the puppet. But the plan is the puppeteer. Or as Drucker suggests, the person-as-subject is produced by the organization of architecture itself.

This inclination can be found in the writings of architects, when they discuss the relationship between architecture and people:

> In human dwelling places, complex *inner* stimuli derive *from* the design of rooms and articles in them with which we surround ourselves. . . .

> It is clearly the design of a room and its furniture which *calls for* certain habitual movements and placements of the body. The taking and holding of a posture, the going into any muscular action, in turn establish . . . kinesthetic pattern.

> The . . . kinesthetic pattern established, inside the body, is . . . in intimate *correspondence with* the layout and design pattern outside. Architecture, in fact, is just such a pattern, laid down about us to *guide* continuously the movement and straining of our eyes, necks, arms, and legs.[13]

In Richard Neutra's extreme conception, even human kinesthetic makeup derives from physical arrangements. Presumably, the joints and structure of the model being permit the way it can move, but architectural form determines its internalized tendencies and the fact of its movement. This practice of turning human beings into architecture has been deepened and extended in Oscar Newman's popular notion of defensible space. Here human consciousness and intersubjectivity are tangential to the strategic arrangement of spaces that can prevent criminal acts and catalyze sentiments of community.[14]

These observations, however, overlook features of architects as people in their own right. They are educated and they accumulate experience. Formally and informally, architects are exposed to theories of society, to literature, to psychology, to ideologies. They have experience with and often participate in policy making that affects the shape the built environment can take. In this respect, we must also ask—even when the connections to design are unclear—what conception of history, of human motivation, of social evolution and social order lies at the base of designers' world views. As Favro shows, notions of social class were as predominant to Vitruvius as kinesthetics were to Neutra. We should explicate the connections, however tenuous, between these world views, the accounts architects give of their activities, and the buildings they design and we live with. To the extent that architects include images of people, we must also inquire about aesthetic intentions.

The Text

Because this book is the first effort to survey the topic, we feel obliged to cover a wide terrain. Our authors mine the social views of architects ranging from Vitruvius to Eisenman, laying out the stances of various historic and contemporary American architects along the way. The contrasting social implications of Modernist and postmodernist stylistic perspectives are drawn. Prevailing cultural and political templates are questioned for their impact on the forms of the house, the hotel, and the high-rise.

We selected the authors because they have something interesting to say on the subject, not because their ideas fit into a neat package of our own design. The chapters provide diverse views about the features of architects' hidden actors, about how we can bring them to light, and about what their influence has been on our surroundings. Each author explores the topic with a particular bias, be it historical, psychological, sociological, architectural theory, or architectural practice, and we have tried to preserve the variety of voices that this multidisciplinary effort produced.

The chapters are organized by four primary themes. The first section, "Ideals in Words," concentrates on the architects' spoken and written images of social life, unfettered by built reality. In the second section, "Shapes of Social Vision," architects' buildings are dismantled to tell the hidden actors' story. "The Subject's Identity," Section III, focuses on the structure, nature, and origins of the architects' people. To close the book, the section titled "Misfunctions and Revisions" examines the widest possible social arena, reinterpreting architecturally related policy, politics, and movements with our imaginary companions in mind. In fact, the reader who wonders why our subject merits book-length treatment will do well to turn first to the very last chapter, a historic overview of architecture's people by Roger Montgomery.

This volume raises questions as it charts new territory. Some architect readers, for example, might ask, "What about the rest of us, who may not be stars but probably designed the buildings where you live and work? Are our people a completely distinct breed?" Although we do not now have an answer to these questions, we think it appropriate that in this first survey, the historical figures and contemporary architects of public note are the ones to draw us into the subject. They exist on the open public record. There are other good questions: Would the social imagings of women architects differ from those of our all-male cast? And never mind the architects' people; what about the developers' people? What do all these people mean for actual inhabitants of architectural space? In response to all these, we have only hypotheses and a great deal of curiosity—curiosity that, if shared by others, we hope will be guided by the present, if incomplete, inquiries to new ones.

Together the book's four sections encompass words and deeds, theory and practice, subject and object. Our objective is to lay out a wide conceptual net that clearly describes the notion of architects' people and its far-reaching implications. Our larger purpose in assembling this collection is to start something. Architecture is an art, a political event, a business, and a humane craft. The first and the last interest us most especially. We hope that, through discussions of the sort included here, architects might develop an informed and artful attitude toward the lives of the people they imagine into their designs: an attitude that might bring ever-new living texture to the art of architecture.

Notes

1. Dolores Hayden, *Redesigning the American Dream* (Cambridge: MIT Press, 1985).
2. The topic appears to have only one short precedent. See Edward W. Wood et al., "Planners' People," *Journal of the American Institute of Planners* (July 1966):278–34. See also Lambert van der Laan and Andries Piersma, "The Image of Man: Paradigmatic Cornerstone in Human Geography," *Annals of the Association of American Geographers,* 72 (1982):411–26.
3. Robert Sommer, *Social Design* (Englewood Cliffs, NJ: Prentice-Hall, 1983), pp. 8–14.
4. The environment and behavior literature is, of course, massive and growing. However, specific attention to architects' social assumptions and intentions is less common. See, for example, Howard Boughey's "Blueprints for Behavior: The Intentions of Architects to Influence Social Action Through Design," unpublished Ph.D. dissertation, Princeton University, 1968; Howard Harris and Alan Lipman, "Architecture and Knowledge: Control or Understanding," *Architecture and Behavior* 1 (1980–81):137–47.
5. The "art defense" was coined by Boughey, "Blueprints for Behavior."
6. Robert Campbell, "Echoes of the Prairie Style on a New England Campus," *Architecture* (October 1985):43.
7. Robert Gutman, in his chapter in this volume, points out that Kahn's inquiry into "what the building wants to be" has been largely misinterpreted as a version of anthropomorphism. He suggests instead that Kahn was referring to the idea that wanted to be realized in the building.
8. W. Russell Ellis, "The Environment of Human Relations: Perspective and Problems," *Journal of Architectural Education* 27 (1974):11–18.
9. Constance Perin, *With Man in Mind* (Cambridge: MIT Press, 1970), p. 116.
10. See, for example, Dana Cuff, "Negotiating Architecture," in *Design Research Interactions,* ed. A. Osterberg, C. Tiernan, and R. Findlay (Ames, Iowa: Proceedings of the Environmental Design Research Association, 1981), pp. 160–71.
11. Joseph Juhasz and John Zeisel, eds., "Social Science in the Design Studio," *Journal of Architectural Education* 34:3 (1981).
12. Alfred Schutz, "Common Sense and Scientific Interpretations of Human Action," in *The Collected Papers of Alfred Schutz,* Vol. 1: "The Problem of Social Reality," ed. M. Natanson, (The Hague: Nijhoff, 1962), p. 41.
13. Richard Neutra, *Survival Through Design* (New York: Oxford University Press, 1954), p. 151 (emphasis added).
14. Oscar Newman, *Defensible Space* (New York: Macmillan, 1972).

IDEALS IN WORDS

To uncover the architects' imagined companions, we can examine two fundamentally different artifacts created by architects: what they say and what they build. In this section architects explain in their own words their visions of person, group, and society. The gift of words, spoken and written, is extended fantasy. For our purposes, architects' words are central because in the imaginal, all forms of life can survive. Architects' words capture an ideal—the way the world ought to be—in a utopian vision that can be fabricated, transformed, purified, and taken to extremes. Although the same might be said of visionary design schemes, actual buildings are far more constraining. Architects' words are not their buildings, but they are windows to the architects' thinking about their designing, even when they are rationales for aesthetic idiosyncrasy: the architectural fact stands out against the foil of the fantasy. Many words about designing are considered here.

When architects speak or write about their work, they are presenting a professional image for public view. That is, they are portraying themselves as they wish to be seen. Perhaps the architects' words are primarily propaganda, but even propaganda can be critically assessed and interpreted. As architects tell their stories of buildings, clients, principles, and methods, we frequently find the elusive population of mental actors lurking between the lines. This is their natural home, since, by nature, the homunculus plays its role from behind the scenes. The critical task, then, is not to express the architect, but carefully to construct his or her imagined companions from the bits and pieces of evidence left by the architect's tale. The architect's words are a text, and as with literature, the responsibility of the critic is to provide a reading that opens the work in a special light.

In the following chapters, the architects do speak for themselves, but in addition, the authors read between the architects' lines to understand the conceptions of society and individuals that have shaped our built environment. These first four chapters raise themes that crop up throughout the book: the architect's self-image, the living building, the client, and the evolution of the homunculus are a few of the ways invented actors can be understood in the architectural context. This section introduces the reader to these themes and others, as well as to methods of reading people into the tales architects tell.

Among the characteristics of the architects' people is their position in society, as Favro points out in the writings of Vitruvius and Alberti. The *ideal* architect or patron of status, education, and wealth has been a silent partner in design from the first century B.C. up to the present day. A utopian social order is apparent in Frank Lloyd Wright's words, having less to do with class per se than with an ideal relation between people and nature. Ellis discusses this Usonian democracy, but he also finds an unsuspected actor among Wright's people: the living building.

Buildings that exhibit human qualities join the architect's mental renderings of the individual and society. The conversation with Joseph Esherick makes us aware that the architect's companions are dynamic and evolving. The snapshots of these invisible actors captured at any particular moment reflect an accumulation over the course of the architect's life. That these images are capable of further development is the premise for a later chapter on the use of literature in design education. In the last chapter, renowned New York architects restate and corroborate many of the ideas mentioned already. Here Cuff explores the relationship between the architect's self-image and the way others are conceived. In the words of Vitruvius and Alberti, as well as of such contemporary architects as Meier and Polshek, we consistently find the architect's own shadow affixed to the notion of who we might be.

This section is not intended as an overview of the historical development of architects' people. Instead, this section and others in the text sample provocative cases in order to explore the conceptual features of mental actors. It is also our aim in this section to understand the ideals of influential architects whose words, as teachers or writers, may have shaped legions of architects' people.

1

Was Man the Measure?

DIANE FAVRO

In the fifth century B.C. the Greek Protagoras wrote, "Man is the measure of all things."[1] From Michelangelo to Le Corbusier, architects have interpreted this famous statement literally, taking the human form as the physical measure of all artistic creation (Figure 1.1), but Protagoras had something different in mind. As a Sophist, his statement reflects a change of philosophical focus from the study of nature to the study of man and his relationships. This is precisely the domain of architecture, the one discipline that deals with social and behavioral as well as aesthetic and pragmatic issues. Yet until the advent of environmental sociology and psychology in the twentieth century, few architectural writers overtly explored theories of human nature and motivation in relation to architecture. These topics simply did not come to the minds of premodern architectural theorists.

Even Vitruvius and Alberti, history's most famous writers on architecture, ignored the homunculus, at least on the surface. Generations have examined their texts for information on the classical orders, proportions, and building forms. In the shadows of these subjects hover the authors' subliminal portraits of the architects' people. The comparison of these portrayals is appropriate. Both Vitruvius and Alberti are highly revered architect-authors. Both operated in eras encased in an elitist, stratified *Weltanschauung* (world view). Both focus on the elite—the aristocratic client and the learned practitioner. For them the architect, or rather the ideal architect, possesses the qualities most highly valued by the society at large. People of lesser status do not matter. Neither Vitruvius nor Alberti devotes much space to those who actually labored on and occupied the final architectural products. Man may have been the measure, but these authors set their calipers to selective scales.

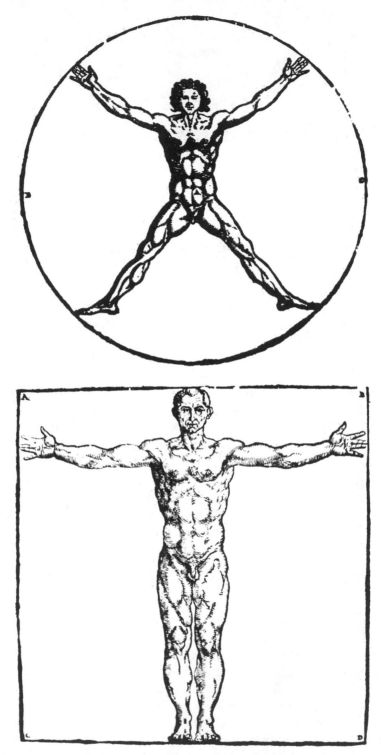

Figure 1.1 Vitruvius related the human figure to the perfect forms of a circle and square. (Illustration for an edition of Vitruvius by Giovanni Antonio Rusconi, 1590.)

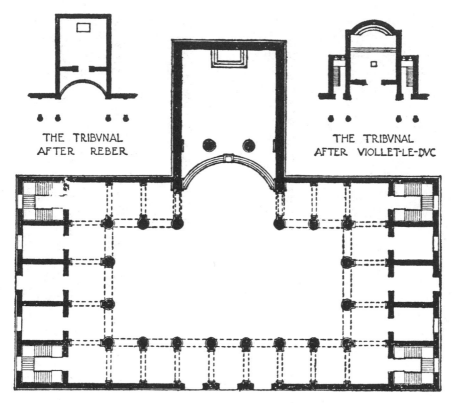

THE TRIBVNAL
AFTER REBER

THE TRIBVNAL
AFTER VIOLLET-LE-DVC

Figure 1.2 Reconstructed plan of the Basilica at Fano designed by Vitruvius. (Morris Hicky Morgan, trans., *Vitruvius: The Ten Books on Architecture* [New York: Dover, 1960], p. 135.)

Vitruvius and *De Architectura*

When Vitruvius settled in Rome at the end of the first century B.C., he was at the end of his career. After years as a military engineer and architect for private and municipal clients, he had much expertise to impart (Figure 1.2).[2] Vitruvius wrote a treatise in ten books on his profession. In content, *De Architectura* was a notable departure from previous architectural writings. Earlier Greek tracts by practitioners generally describe a single building. Those by nonarchitects (e.g., Plato and Aristotle) deal with the built environment and its inhabitants in order to support philosophical arguments. In contrast, Vitruvius' work outlines the education and duties of the architect, the client's role, and contemporary architectural principles. *De Architectura* was one of the earliest treatises by a *practitioner* to explore the architect's world.

Vitruvius' treatise is at once practical and promotional. The ten books deal with construction techniques, basic architectural principles, siting, building types, materials, proportions, and acoustics, yet *De Architectura* is not just a handbook. On returning to Rome after a lifetime of service, Vitruvius found architects occupied a low rung on the social ladder. In reaction, he wrote to promote the architect's status.

The negative image of late Republic architects was in part deserved. With shame, Vitruvius admits it was common practice for contemporary practitioners to grovel and misquote the budget in order to get commissions. He sadly acknowledges that educated Romans were not attracted to a field lacking an ethical code and standards of excellence (10.2). As a result, many projects were undertaken by uneducated, unrecognized builders who produced low-quality buildings, which, in turn, further denigrated the profession.

In Roman society, the highest goal was immortality through enduring recognition. An individual's importance derived from family status, education, skills, and above all, acquired authority, or *auctoritas*. Equated with "influence," "power," and "prestige," *auctoritas* ensured remembrance by posterity. Through honorable deeds, grand undertakings, and good character an individual garnered this valuable but elusive commodity. The more *auctoritas* accrued, the more a man was honored in life and remembered after death. In special cases, this accolade was awarded to inanimate objects. Vitruvius praises Augustus by acknowledging the emperor's buildings had their own *auctoritas* (1.2).

The practitioner, the client, the builder, the user, and the structure itself are the primary actors on any architectural stage. Vitruvius recognized that of these cast members, only the aristocratic client and the outstanding structure merited *auctoritas*. With *De Architectura*, he persuasively argues for the inclusion of another individual, the architect. To strengthen his case, he minimizes the role of all the architects' people incapable of earning *auctoritas*.

How could the architect lay claim to his own *auctoritas?* In the hierarchical society of the first century B.C., wealth could bring a practitioner comfort and notoriety; it did not necessarily bring respect and enduring memory. Design distinctiveness, too, did not engender *auctoritas*, for Roman designers all operated within the same well-defined, traditional framework. Even the creation of recognizably great buildings brought the architect only a modicum of fame; a building's *auctoritas* accrued to the patron. Although today we hear of Graves' Portland building or Johnson's AT&T building, in the first century B.C., the Roman spoke of the Baths of Agrippa or the Basilica of Julius Caesar. Only rarely was the name of an architect linked with a building.[3] The practitioners respected, discussed, and admired by Roman society belonged to a very select group—those who associated with famous patrons or wrote treatises.

Vitruvius advises architects to bask in the reflected light of important patrons. Referring to painters and sculptors, he explains, "Those . . . [who] have come down to posterity with a name that will last forever . . . acquired it by the execution of works for great states or for kings or for citizens of rank . . . [those] who executed no less perfectly finished works for citizens of low station, are unremembered" (3.2).

Hungry to gain *auctoritas* for himself, Vitruvius actively cultivated every possible link with the imperial family and dedicated his treatise to Augustus. Although apparently not very successful at networking in the imperial circle, Vitruvius learned from the experience.[4] In *De Architectura*, he assigns the architect the personality traits of a typical Roman gentleman, or *ingenuus*. Both are decorous, courteous, incorruptible, rational, clever, refined, manly in spirit, and educated. Vitruvius holds himself up as an exemplar, stating, "Other architects beg and wrangle to obtain commissions; but I have been taught by my masters that it is the proper thing to accept a commission only after being asked, and not to ask for it; since a gentleman will blush with shame at petitioning for a thing that arouses suspicion" (6.5).

For the Romans, social graces and high morals were irrevocably linked with knowledge. Reasoning was the domain of the higher social classes. Knowledge, or the veneer of knowledge, made almost anything possible. Vitruvius wistfully notes that in the venerated past, architects were chosen first for their family status and second for their education. He goes on to praise patricians with no design training who, "in the confidence of learning," design their own residences (6.6). Vitruvius urges the architect to be proficient in history, law, music, and literature, that is, in all learned subjects. He explains that without a liberal education no architect, regardless of his competence, would ever reach a position of importance, would ever earn *auctoritas* (1.1.2).

Further, Vitruvius simply states, "an architect ought to be an educated man so as to leave a more lasting remembrance in his treatises" (1.1.4). Authorship of a written work brought enduring renown, as Vitruvius well knew; the majority of Greek architects he mentions in *De Architectura* had written treatises. Bemoaning the lack of texts by Roman practitioners, Vitruvius wrote to acquire his own *auctoritas* (7.15). Throughout the ten books, he presents himself in idealized form. He emphasizes his own broad education and whenever possible names famous contemporaries whom he has met. But it is the very act of writing that elevates Vitruvius. He explains, "thinking it beneath me to engage with the uneducated in the strug-

gle for honor, I prefer to show the excellence of our department of knowledge by the publication of this treatise" (3.3).

Since the status of any practitioner relates to that of the entire discipline,
Vitruvius promotes architecture as a learned undertaking based on rational
principles. He extols the architect to consider issues and ideas, to analyze
how a building project simultaneously deals with *firmitas*, *utilitas*, and
venustas ("stability," "function," and "beauty;" 1.3.2). In addition, he links
architecture with the rhetorical arts, calling it a "magnificent discipline" on
a par with the liberal arts of medicine, astronomy, and oratory (6.6). In
fact, Vitruvius elevates architecture above other studies, stressing the difficulty of mastering this learned discipline: "I think that men have no right
to profess themselves architects hastily, without having climbed from boy-
hood the steps of these studies, and thus, nursed by the knowledge of many
arts and sciences, reached the temple of architecture at the top" (1.1.11).

The Roman author devotes the first of his ten books to the ideal education for an architect.[5] With calculated care, he tells how book learning can
serve the practitioner. He explains that an understanding of medicine enables the architect to design structures promoting good health, familiarity
with music allows him to work with proportions, knowledge of law permits
him to foresee possible liabilities, and so on. Significantly, Vitruvius does
not overemphasize applied knowledge. As always, he draws the distinction
between the educated elite, including architects, and the uneducated lower
classes. He argues that although even an ignorant layman, *idiota*, can recognize quality architecture, only the learned architect is able to *concep-
tualize* an unbuilt structure's appearance (6.8.10).

For all his discussion of book learning, Vitruvius could not ignore the
value of hands-on experience. After many years of actual practice, he knew
that those who relied "only upon theories and scholarship were obviously
hunting the shadow, not the substance" (1.1.2). Succinctly, he argues that
an *architectus* needs both theory and practice (1.1.15). Yet actual practice
brought the Roman architect into contact with masons, bricklayers, wood-
workers, slaves, plebian users, middlemen, and other undistinguished individuals. For Vitruvius, these were not the right architects' people.

Vitruvius describes a sophisticated, erudite practitioner more comfortable
in a private salon than at a construction site. The author's own profile was
removed from the ideal. After spending a lifetime as a military engineer-
architect, Vitruvius lacked finesse; he probably did not converse easily with
Augustan court members, if he ever saw any. Furthermore, it had been
many years since his schooling. Vitruvius' Latin is inelegant, his familiarity
with learned writings superficial. Although he cites virtually all important
Greek philosophers, he hardly distinguishes between the theories of each.

For this architect-author, references to history and posterity helped define the practitioner as a learned man; specific theoretical content was of secondary import.

Vitruvius relied on revered Greek texts to identify subjects of import to architects. For example, since building utility was not explored by Greek authors, Vitruvius does not examine this topic in depth. In contrast, he expounds on the optical studies of Hellenistic researchers, carefully explaining how the architect can adjust a structure to counteract the distortive tendencies of the human eye (3.5.9; Figure 1.3).[6] This selectiveness establishes a distinct hierarchy. Readers infer that perception belongs to a higher order than use. More important, they realize the learned architect has the skill to manipulate the individual perceiver and to overcome the limits of nature by fooling the eye.

Vitruvius further demonstrates how the architect can use his knowledge to manipulate nature. He advises the practitioner to study regional variations in architectural requirements and in human personalities. He explains, "the effects [of climate] . . . are observable in the limbs and bodies [and character] of entire races" (6.1.3). Chauvinistically, Vitruvius identifies Italy's climate as the most ideal in the Mediterranean, and the Latin race as the most perfect in body and mind. He notes, "southern nations have the keenest wits, and are infinitely clever in forming schemes, yet the moment it comes to displaying valour, they succumb because all manliness of spirit is sucked out of them by the sun" (6.1.10–11). Vitruvius explains how the skilled architect can design to compensate for undesirable regional conditions and traits. Thus, he implies the practitioner can approximate the environment of Italy and stimulate building occupants to develop the preferred "Italian" personality.

Vitruvius' architect treats anonymous users like any other inanimate determinant of architectural form. He considers people separately from buildings. Thus, the author separates his animated description of southerners from his discussion of the houses they occupied. Generally, Vitruvius peoples architecture only when clarifying the configuration of an imported building type or when describing a variation. For example, in book five he mentions athletes in relation to the uniquely Greek running track and actors in relation to the differences between Greek and Roman theaters. In the passage on his own design of the basilica at Fano, Vitruvius carefully explains where different users stood in order to justify his deviation from the traditional basilical form (see Figure 1.2).

Furthermore, in discussions relating to the people who use buildings, Vitruvius structures his sentences with architectural features as the primary nouns; people appear in the genitive tense. A room used by painters is a

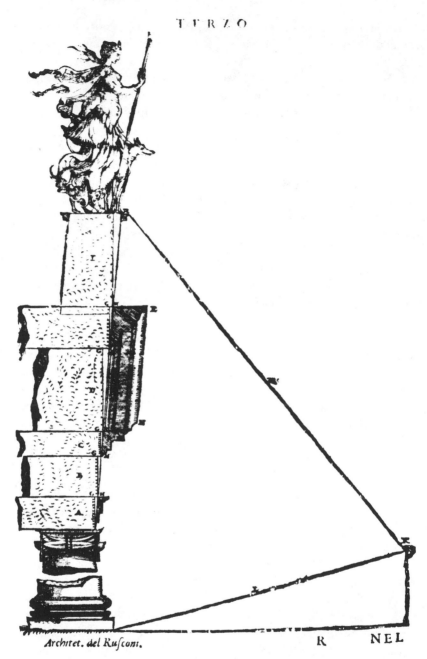

Figure 1.3 Vitruvius recommended entablatures be adjusted optically to compensate for visual distortion. (Rusconi.)

"painters' studio" (6.4.2). Similarly, when explaining *firmitas, utilitas,* and *venustas,* Vitruvius uses the passive tense, thereby stressing content and application, not how these issues affect human actors. This treatment places architecture in the primary position, with the human user in a removed, secondary role.

The consideration of people as objects is intensified by the use of humans as physical models and metaphors. Taking Protagoras' statement literally, Vitruvius describes how a well-formed human body reflects the most perfect geometrical shapes: the circle and the square (3.1; see Figure 1.1). Like these elemental shapes, the body displays unity, symmetry, and formal integrity. Architecture should do the same. Vitruvius advises the architect to design houses with public spaces and intimate private areas displaying uniform grandeur; he explains, "there will be no propriety in the spectacle of an elegant interior approached by a low, mean entrance" (1.2.6).

Vitruvius also explores direct formal interrelationships. On the most basic level, he notes that the use of body parts as standards of measurement irrevocably links people and architecture. Further, Vitruvius narrates Greek legends about the anthropomorphization of the column. The Doric shaft has the proportions of a man, the Ionic those of a matron, and the Corinthian, those of a maiden (4.1; Figure 1.4). He advises the architect to use the manly Doric order for the temple of a virile god like Mars or of goddesses in their masculine guise (1.2.5). Although he associates proportional systems and personality traits, this connection too seems oddly dehumanized; the emphasis is on standardized formal characteristics, not enlivened human traits.

When relating personality traits and architecture, Vitruvius focuses on those who possess *auctoritas:* specifically, gods and upper-class clients. He charges the architect to make the status of these important groups evident in every aspect of their physical environment, especially in their residences. He calls for temples and private homes to be the physical embodiment of *auctoritas.* Vitruvius devotes all of book six to the single-family urban house of the patrician class. He identifies the patron of such a structure as the *dominus,* a word associated not only with the verb *dominor* ("to rule") but with the single-family structure itself, the *domus.*

In discussing the *domus,* Vitruvius admits that functional differences and tradition naturally affect architectural layout, but he stresses that image remains of paramount importance. He recommends, "persons of high rank . . . should be provided princely vestibules, lofty halls . . . libraries and basilicas . . . because in such houses public deliberations and private trials and judgments are often held" (6.5.2). Though this passage links actual activities and architectural settings, the author connects not only form and

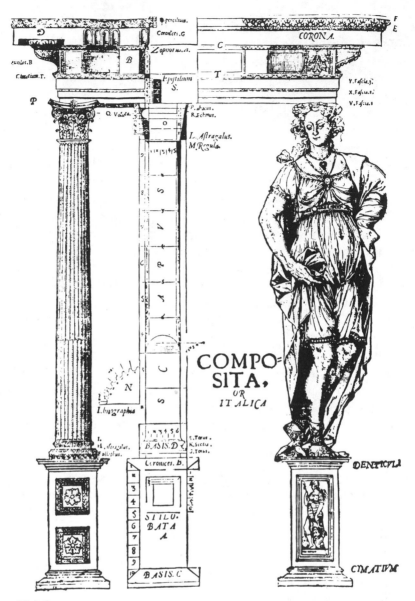

III-A-7.

Figure 1.4 Vitruvius related the proportions of the column to those of a human being and recorded the origin of columnar caryatid forms. (Illustration for an edition of Vitruvius by John Shute, 1563; reprinted with permission of Gregg Publishing Co. Ltd.)

function, but also form and status. More specifically, Vitruvius associates the learned discipline of architecture with Rome's most powerful citizens, those who conducted formal audiences, read books, and held courts within their homes.

In contrast, the author of *De Architectura* barely mentions the apartment residences occupied by the middle and lower classes of contemporary Rome. Multistoried, pragmatic *insulae* lacked history, prior learned analysis, and associations with important individuals. Within the same unprestigious architectural category Vitruvius groups gigantic private warehouses and other revenue-producing constructions. Such mundane projects were the products of anonymous builders and speculators and were the workplaces of unimportant plebians, the nonpeople as Groth calls them in a later chapter. These structures could not earn an architect *auctoritas*; they did not belong in a learned treatise.[7] Only the most traditional, the most conservative building types have a history; only they are worthy of discussion.

According to Vitruvius, neither patrons nor architects could earn *auctoritas* if they varied widely from established traditions of dignified conduct or from accepted buildings forms. Each seeking personal fame, patron and architect at times came into conflict. The author notes that the client and building receive praise when opulent materials are used, yet the architect acquires repute by exercising restraint and limiting expenses (6.8.9). To minimize clashes, Vitruvius outlines the appropriate spheres of action for the practitioner and recommends an elevated standard of conduct. For the client, he presents guidelines for judging existing and planned projects.

Outside the treatise, reality took its toll. The architect might long to converse with the client as an equal and design monuments; more often practitioners, including the author of *De Architectura*, ingratiated themselves with powerful patrons and earned a living by designing warehouses or engineering projects. Clients, in turn, struggled to balance the elevated search for *auctoritas* with the desire for profitable architectural investments. Vitruvius wrote about himself as the model professional. He did not provide an equally lengthy description of the exemplary client.

For information on Roman patrons, another author must be consulted. Cicero—senator, orator, and moralist—dealt with architectural patronage in his letters and learned texts. In *De Officiis* (ca. 44 B.C.), a late work on moral philosophy, he describes the proper residence for a man of rank and station. To start, Cicero states that the prime purpose of a *domus* is serviceability, but like Vitruvius, he immediately moves on to issues of more magnitude: image and status. Viewing function in terms of social rank, he explains that social stature is reflected in the number of guests a patrician could comfortably entertain or receive; "if it is not frequented by visitors, if it has an air of lonesomeness, a spacious *domus* often becomes a discredit to its owner (*De Off.* 1.39). Conversely, a house of proper form and elegance could be an asset. Cicero describes an imposing *domus* in Rome, noting that, "Everybody went to see it, and it was thought to have gained votes for the owner" (*De Off.* 1.39). As a moralist, Cicero does at one point advocate restraint in building, yet after his previous discourse on the fame-inducing capabilities of opulent homes, few would be swayed.

In *De Officiis*, Cicero is not concerned with the architect. He lists architecture as one of the professions requiring higher intelligence, but does not discuss the practitioner at length (1.42). From Cicero's perspective, the influential owner and the building itself were the most important of the architects' actors. Yet one was superior. He argues, "a man's dignity may be enhanced by the house he lives in, but not wholly secured by it: the owner should bring honour to his house, not the house to its owner" (*De Off.* 1.39).

Cicero portrays architectural patronage somewhat differently in his letters to friends and relations. The more relaxed literary form allowed him to deal with topics inappropriate for erudite publications. Here he criticizes an inept architect-builder, Diphilus, who could not use even a plumb-line correctly. Corresponding with his familiars, Cicero writes about assertive clients addicted to real estate speculation. Late-Republican patricians acquired new buildings as readily as new togas. Although not considered wealthy, Cicero himself owned over twenty structures.[8] In letters, he freely admits to owning risky, high-profit/low-status investment property and openly displays his concern with profitability as well as social status and appearances. Cicero's writings reflect the split personality of the Roman patrician class. One side focused on fame, abstracted learning, and morality; the other, on money, politics, and investments.

In part, the duality between the written world and the real world resulted from the Roman authors' elitist definition of the architects' people. Both Cicero and Vitruvius addressed the same readers—wealthy, learned patricians; as a result, they focused upon a selective cast of characters. Laborers and low-class users remained behind the scenes. Yet through a "trickle-down" effect they too benefited from the written image of the architect. By promoting the profession as a learned discipline, Vitruvius helped stimulate general interest in architecture. As wealthy clients focused on grand, well-ordered structures, they improved the city's environment for all, even anonymous users. Similarly, as architects became concerned with a broad range of topics, from medicine to law, they produced healthier, better organized, more solid structures. Builders, as well as users, profited.

In the unsettled environment of the first century B.C., Roman society began to break out of its tradition-bound, hierarchical restraints. Individuals were still ranked according to their ethnic group, clan, social class, and education. However, the ambitious sought ways to improve their standing. Vitruvius urges architects to exploit social mobility, outlining how a practitioner could transform himself from an obscure builder into a revered professional through education, good social contacts, and promotion of the discipline. Simply put, the author laid out a way for the architect to transcend his traditional position in society by creating a new one. The successful architect basked in the reflected fame of his buildings and clients; even better, he earned long-lasting glory for himself and his profession by writing. Given these calipers, Vitruvius did indeed measure up.

Today the goals Vitruvius set for his profession largely have been realized. Architecture is considered a learned discipline taught in universities; practitioners are held in esteem; buildings are named after architects as well as clients. Vitruvius himself is recognized as a great man. This transformation was extremely slow. *De Architectura* did not immediately improve the status of either Vitruvius or the profession. Except for his own text, nothing indicates Vitruvius was highly regarded in the Augustan era. If the treatise had brought immediate success, other architects would have scrambled to publish their own works. Few did. Furthermore, *De Architectura* did not significantly improve the status of the architect in Roman society. Not many men of standing entered the profession. Martial, writing a generation after Vitruvius, advised a worried Roman father, "if your son seem to be of dull intellect, make him an auctioneer or architect" (*Epigrams* 5.56).

In following centuries, the existence of *De Architectura*, more than its contents, brought Vitruvius fame. His was the only Roman architectural treatise to survive classical antiquity. Each subsequent age dissected the text searching for information both to document the revered past and to support contemporary beliefs.[9] During the early Renaissance, humanists "rediscovered" the ten books on architecture and began to draw upon the contents for information on classical environments and concepts. In particular, they were attracted to the Roman notion of the individual as a driving force in society. Inspired by the work of Vitruvius, architects and patrons of the Quattrocento began to use buildings and writings about architecture to define themselves.

Alberti: *De Re Aedificatoria*

In the mid-fifteenth century, a humanist suggested to Leon Battista Alberti (1404–1472) that he write his own version of Vitruvius' treatise.[10] The illegitimate son of an exiled Florentine, Alberti had studied Latin and Greek at Padua and law at Bologna; in the 1440s, he held a papal post that allowed him the leisure for research. With scholarly concern he set upon the task of updating the Roman text. Alberti had already written respected books on moral and social subjects as well as on cartography, painting, sculpture, and the antiquities of Rome.[11] In his works, he drew upon the traditions of both the ancient and recent past and advocated new ways of perceiving the world.

Alberti presented his work, *De Re Aedificatoria*, to Pope Nicholas V in 1452, probably in rudimentary form. The final Latin text was published posthumously in 1485. Other Renaissance authors had been content to reprint Vitruvius's original work with new commentaries and illustrations. Alberti was not. He drew inspiration from the Roman source, but grounded the new treatise in the theories of his own day, creating a Quattrocento equivalent. Even in its title, Alberti's work reflects this derivative but distinct association with Vitruvius' *De Architectura*. Like the Roman author, Alberti arranged his work in ten books. Further, he loosely organized the material around the three Vitruvian essentials: *utilitas, firmitas,* and *venustas.* However, in content and emphasis, Alberti diverged from his model. *De Re Aedificatoria* was a tour de force of humanist learning, reflecting the author's vast knowledge of ancient sources, his own studies of ancient buildings, his technical knowledge based on research and personal experience, and his mental images of individuals and society.

In the fifteenth century, Italian society was as hierarchical and status conscious as that of ancient Rome. With the growth of powerful, competing city-states came ambitious individuals protective of their right to self-determination. Quattrocento artists of all kinds struggled to distance themselves from the restrictive medieval guild system. Alberti led the fight. In *Della Pittura*, from the 1430s, he advises painters not to restrict themselves by association with a guild, but rather to cultivate individuality as apprentices under a master painter. He emphasizes the personal bond between the individual patron and the individual artist rather than the more alienated interaction between learned advisers and guild members. Above all, Alberti urges the architect to disassociate himself from all groups and operate as an individual.

Individuality fed a desire for personal repute. Fame, in large part, relied on *virtu* and *magnificentia*. Roughly parallel to Roman *auctoritas, virtu* was equated with power, honor, morality, and virtue; *magnificentia* was associated with loftiness of thought and action.[12] Both brought enduring remembrance. For Alberti, just as for the architects of the Roman past, glory was measured by memory; the opinion of future observers was as important as contemporary judgments. In *De Re Aedificatoria*, he calls on the practitioner to reflect always, "what manner of man he would be thought . . . how much applause, profit, favour, and fame among posterity he will gain when he executes his work as he ought, [and] if he goes about anything . . . unadvisedly . . . to how much disgrace, to how much indignation he exposes himself . . . ." Alberti explains, the architect labors above all "to deliver his name with reputation down to posterity" (IX.x).

The need for recognition was obvious. Architects of the early Renaissance, like their Roman predecessors, did not share in the fame of their designs as much as the patrons whose names or emblems boastfully appeared on building façades (Figures 1.5 and 1.6). Alberti's writings champion the importance of the architect. In his allegorical satire *Momus* (1443–52), he presents Jove as an appropriate judge of the architect's accomplishments. The father of all the gods praises architects as the only persons capable of redesigning the cosmos.[13] Yet such godly praise was not enough; the Renaissance architect needed human approbation.

Alberti defines the architect's human evaluators as groups: the learned, the generous, those of public spirit and means (pref.). He does not include the uneducated general public. Alberti records that if beauty and ornament are well executed, "there is hardly any man so melancholy or stupid, so rough or unpolished, but what is very much pleased. . . ." Yet he acknowledges that it was a "common thing with the ignorant, to despise what they do not understand" (VI.ii). He recommends that the architect seek out patrons of highest rank and quality, especially those who love the arts. Only the learned could rightly judge, and rightly appreciate the cerebral aspects of Renaissance architecture. Similarly, only the generous could afford and appreciate great expenditures; only those with public spirit could recognize architecture's potential for social good.

Alberti himself wrote for aristocratic audiences and mingled with the most famous men of his time, including Pope Nicholas V, Frederigo da Montefeltre, and Giovanni Rucellai. Lesser patrons would detract from the architect's *virtu*. In book nine, Alberti argues, "Why should I offer those inventions which have cost me so much study and pains to . . . persons of no taste or skill?" He warns, "work loses its dignity by being done for mean persons . . . the authority of great men . . . advance[s] the reputation of those who are employed by them."

As Vitruvius before him, Alberti focuses on the patron and the practitioner as the most important of the architects' people. However, his portrayal of the practitioner differs from that of the ancient author. Looking in the mirror, he saw a very different reflection. Vitruvius wrote his treatise as an older practitioner, anxious to determine ways to rise in society; unrefined

Figure 1.5 Palazzo Rucellai in Florence, by Alberti, 1446–51.

Figure 1.6 Façade of Santa Maria Novella in Florence, designed by Alberti, 1448–70.

himself, he urged the architect to cultivate social graces. Alberti, as a cleric and noble, moved in elevated circles long before he turned to architecture. In many ways, the Quattrocento author filled Vitruvius' description of the ideal architect. He had a solid education in the liberal arts, he belonged to the proper upper class (albeit as the illegitimate son of a nobleman), and he had gained fame through writing. Though impoverished, Alberti achieved a high social position on the basis of his personal knowledge and independent scholarly achievements. He was a product of his own self-determination. Alberti's favorite motto echoed contemporary sentiment: "men can do anything they want."[14]

In *De Re Aedificatoria*, Alberti describes the ideal architect as likewise a self-made man, an individual forged by the fires of his own making. He repeatedly calls on the practitioner to mold his own future, "to produce something admirable, which may be entirely of his own invention . . ." (IX.x). With this emphasis on self-determination, Alberti approaches learning not merely as a means of helping the practitioner to converse with learned clients or to create an admirable structure. The architect seeks knowledge and Reason for himself.

Renaissance humanists felt Reason ordered the world, making even nature understandable.[15] In *De Re Aedificatoria*, Alberti isolates the essential logic in architecture by distilling all actions and products to pure Reason. In fact, he mentions Reason so often that it can be added to the architects' people. Where Vitruvius instructs the architect to identify formal categories by examining written texts and historical buildings, Alberti challenges him to isolate the abstract commonalities inherent in such works. He makes a clear distinction between personal taste and rational judgment and advises the architect to use nude, simple presentations in his drawings so that the viewer will admire his mind, not his hand.

Alberti also relies on architecture's theoretical side to elevate the profession as a whole. He calls upon all "honest and studious mind[s]" to accept their responsibility "to free this science, for which the most learned among the ancients had always a very great esteem, from its present ruin and oppression" (VI.i). Alberti criticizes "a certain author" (i.e., Vitruvius) for expecting the architect to be as skilled as lawyers, astronomers, and musicians in their various disciplines (IX.x). He presents a more reasonable curriculum, one suspiciously close to his own generalist training. He associates architecture not with the rhetorical arts, but with the four "higher" sciences of geometry, astronomy, arithmetic, and music, the quadrivium requiring theoretical analysis.[16]

For Alberti the discipline belongs in the theorem-oriented, provable sphere of the scientist, not in the ad hoc world of the craftsman. His own examination of the subject was scientifically based, using the same meticulous research methods he earlier applied to other subjects.[17] Alberti calls for objective analysis; he recommends the architect "wait a while until your enthusiasm for your project has boiled over. Afterwards, you may return and consider it more carefully, when your judgment is no longer swayed by fondness for your design but guided by calm Reason" (II.ii).

Before he began to study architecture, Alberti had little architectural experience. His early forays into the practice approximate scientific experiments. For example, while writing his architectural treatise, the cleric was prior at San Martino, Gangalandi; the church's redesigned apse has been attributed to him. With its pilasters leaning to compensate for optical distortion, this project has the feel of a learned experiment (Figure 1.7).[18] Alberti carefully records and evaluates other experiments. He confesses, "when I have come to reduce [my ideas] into lines, I have found in the ones which most pleased me, many gross errors . . . and measuring every

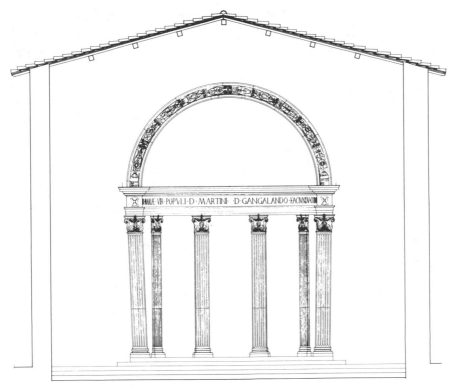

Figure 1.7 Apse of San Martino in Gangalandi, by Alberti, ca. 1432. (Franco Borsi, *Leon Battista Alberti* [Milan: Electra Editrice, 1977], p. 288.)

part by numbers, I have been sensible and ashamed of my own inaccuracy" (IX.x).

Writing before he designed any large projects, Alberti emphasized the architect's ideas, not the physical object he creates.[19] Throughout his treatise, he minimized the role of buildings. Contemporary structures stand backstage, dismissed as the "whims of the moderns." (VI.i)[20] Only ancient buildings of "tolerable reputation" stand out as important actors that mime the universal order of nature (IX.x). Alberti proffers these structures as specimens, as embodiments of concepts, purified from any association with use.

For Alberti the use and hence the users of buildings are not as interesting or important as abstract principles. He looks at individuals in relation to the cosmos, not in relation to other people or to architectural form. When examining ancient buildings, he only briefly considers the people who constructed or used them; "I believe . . . in those who built the thermae, the Pantheon, and all those great works . . . [but] I believe much more in

Reason than I do in any person."[21] The negation of social issues in favor of cerebral topics is understandable in the writing of a scholar. Yet why would the man who wrote an entire treatise on family relations minimize interpersonal exchanges?

The targeted audiences of Alberti's works provide a partial answer to this quandary. In his moralistic tracts the author addressed a broad lay readership; for this audience Alberti wrote in the Tuscan dialect. In his learned treatises he addressed a more elite group, explaining, "I do not write only for the use of workmen, but for all such as are studious . . ." (II.xi). Alberti addressed this audience in Latin. With pride, he points out that the Latin in *De Re Aedificatoria* is far superior to that in *De Architectura*, implying his work should be considered more purely Roman than the Roman's (VI.i).[22]

By writing in Latin, Alberti distances the discourse on architecture from the shops or building sites of the workers and locates it in the salons and studies of the well-educated, discerning elite. Alberti envisions the architect as a learned man on a par with a learned patron. Both educated, these two actors appeciate each other's ideas and together seek perfect form, balanced by an understanding of Reason. In pursuit of Reason, the learned architect (i.e., Alberti) would willingly clamber over ancient buildings, measuring and touching the dirty stones; he would never do the same at a contemporary work site.

During the medieval period, morality had constrained architectural patronage; men were measured by their poverty and good works rather than by showy buildings.[23] As the political and economic environment changed in the Quattrocento, physical manifestations of individual achievement became desirable. Alberti uses his writings to associate building with personal fame. Addressing a broad audience in *Della Famiglia*, he explains, "at times it is proper to incur certain expenses for the honor and reputation of a family like ours . . . I am referring particularly to spending money for the construction of public buildings."[24] Addressing the aristocracy in *De Re Aedificatoria*, he becomes more forceful, arguing that the educated, wealthy elite are innately predisposed to become involved with architecture; "nobody who has the means, but what has an inclination to be building something" (pref.). Like the ancients, he believes the aristocracy should value ostentatious architectural displays as explicit reflections of an admirable moral system. With homage to Cicero, Alberti writes, "Because we decorate our house as much to adorn our fatherland and family as for the sake of elegance, who will deny that such activity is the duty of a good man?" (IX.i).

Alberti's patrons are true lovers of the arts. The author observes, "the meaner sort build only for necessity; but the rich for pleasure and delight" and fame (V.xiv). He identifies two primary ways architecture promotes the patrons' status. First, good design, large scale, and opulent materials overtly advertise *virtu* and *magnificentia* (II.iii). Second, well-executed buildings provide an appropriate stage for dignified actions. Alberti explains that good architecture "enables us to lead a dignified, pleasant life" (I.vi). The "us" is obvious—those who could read the treatise, discuss philosophical issues, and afford great constructs.

Alberti acknowledges that wealth is important; only the wealthy have the resources to purchase high-quality materials and to see their projects to completion. Familiar with the many unfinished buildings of his day, he warns the architect to consider the patron's financial standing (II.i). An incomplete building does not bring honor, nor does one poorly executed by low-paid, incompetent workmen or one left derelict when finished (X.i). To acquire *virtu*, the architect must work with clients who are both learned and wealthy. Although informative, Alberti's characterization of patrons is impersonal. These powerful individuals do not come to life. They recede in importance before the lengthy descriptions of the educated architect.

Alberti charges mankind, and especially wealthy patrons, to be cognizant of the obligation owed to architects for solving pragmatic problems of shelter, engineering, and defense. Moreover, he credits the architect with affecting the health, mental state, business, and status of "us," the elite readers; "we are exceedingly obliged to the architect; to whom, in time of leisure, we are indebted for tranquility, pleasure and health, in time of business for assistance and profit; and in both for security and dignity" (pref.). Alberti tempers this laudatory portrayal by humanizing the architect. For example, he admits the practitioner can "be so carried away by the desire of glory as to rashly attempt anything entirely new and unusual" (IX.xi).

Like Vitruvius, Alberti distances his architect from the mundane. He recognizes an architect should have practical training, yet downplays this admission. When listing the prerequisite skills for an architect, Alberti buries construction experience amid a long list of *intangible* faculties, including wit, enthusiasm, prudence, deliberation, wisdom, and above all, sound judgment (IX.x). He divides architecture into duties appropriate either to the mind or to the hand. Those in the first group have the most positive associations; here the learned architect reigns supreme. In the second group Alberti places physical endeavors (such as brick laying and carving) and construction management, activities undertaken by common workers and overseers, people with no hope of acquiring *magnificentia*.

Alberti clearly describes his image of the laborer. He writes, "it is not a carpenter or a joiner that I thus rank with the greatest masters in other sciences; the manual operator being no more than an instrument to the architect" (pref.). Later in his text, Alberti warns the architect about possible defects of this 'tool'; "remember how difficult it is to find workmen that shall exactly execute any extraordinary idea which you may form . . ." (IX.xi). Significantly, Alberti assigns the builder those responsibilities rejected by the architect. "To run up anything that is immediately necessary for any particular purpose, and about which there is no doubt of what sort it should be, or of the ability of the owner to afford it is not so much the business of an architect as of a common workman" (IX.x). For Alberti, any uneducated builder can satisfy particular needs, but if there is a question as to building form or affordability, then an architect is needed to deduce the answer. The educated architect alone can use Reason in situations of doubt.

Between the revered architect and the lowly worker, Alberti places an intermediary, the overseer (IX.xi). He describes the ideal overseer in generic terms as honest, diligent, and severe, a useful "Tool" (IX.xi).[25] Alberti knew from observance, and later from his own experience, that contemporary practitioners often faced cost overruns, cheating contractors, rigged competitions, inflexible guilds, defaulted payments, and disrespect.[26] He offers the overseer as a buffer that protects the architect from mundane concerns.

Alberti likewise treats architecture's anonymous users as commodities, not as self-determining individuals. When he recommends that the architect provide carefully planned room adjacencies and comfortable spaces in rural estates, he does so not to serve the laborers' comfort, but to improve overall efficiency for the villa owner (V.xv). In his chapters on the city, Alberti treats occupants and their activities as municipal assets or liabilities. For example, since unproductive, unsightly derelicts detract from a city's stature, Alberti urges ruling princes to order cripples and other undesirables to work at a trade in order to stay in the city. They should place those who cannot work in hospitals out of public view; "by this means . . . the city [is] not offended by miserable and filthy objects" (V.viii).

For the learned author, utility and structure concern the architect not because they make a building or city more comfortable, but because "if neglected they destroy all the beauty and ornament" (IX.ix). He charges the architect with providing "fit and suitable conveniencies for every rank and degree of [people], as well masters as servants, citizens as rustics, inmates as visitants" (IX.ix). To distinguish the structures associated with important individuals, Alberti advises the architect to employ a grand scale, rich materials, and particular building forms. He explains that the elemental trabeated system is appropriate for temples (i.e., churches) and the residences of influential individuals. For average citizens, he recommends that the structures of the aristocracy by downscaled in proportion to the users' status.

These buildings are to be identified by arcuated porticos (IX.iv).[27] Architecture for the lowest social classes merits little consideration; Alberti does not discuss them in depth.

In the Quattrocento, the exact layout of buildings was determined by tradition. Alberti explains, "deviating from established custom generally robs a thing of its whole beauty, as conforming to it is applauded and attended with success" (I.ix). The architect's task is to refine existing configurations. Alberti recommends that the practitioner turn to nature for examples that balance functionality and beauty. He writes, "an edifice is a kind of body" to which nothing could be added or subtracted but for the worse (pref.). Interestingly, in book six Alberti holds up the body of a horse, not of a human, as an exemplar (iii). For him, human superiority lies not solely with the body's functional, balanced design but with the ability of the mind to recognize proportionality and appropriateness in all things (IX.x).

Alberti also draws a parallel between architectural forms and human traits. Whereas Vitruvius associates buildings and specific races, the Renaissance author extends the analysis to personalities. For example, he lists the tyrant's need for protection, isolation, and surveillance and gives the proper architectural response—a castle set apart from other buildings, replete with observation towers and pipes for eavesdropping (V.iii–iv). With this simple articulation of need and solution the author draws a concise portrait of the tyrant's character. For Alberti, contact with a building should evoke the same response as contact with the patron. Thus, regarding churches he writes. "I would have every part so contrived and adorned, as to fill the beholders with awe and amazement . . . and almost force them to cry out with astonishment" (VII.iii).

Alberti goes on to explain how the learned practitioner should use architecture to mold individual behavior and emotions. In designing for the clergy, he urges the architect to create model environments that provide neither the opportunity nor the inclination to be unchaste (V.vii). Further, he calls on the architect to accept responsibility for society's communal health. It is the practitioner's duty to enhance the welfare of the patron, which in turn promotes that of the family and of the state (pref., IX.i). Alberti exhorts the architect to consider how building parts and individual structures interact for the betterment of all. A good building, like a good citizen, should not be falsely raised above its proper station either symbolically or literally; it should not induce envy (IX.i–ii). Thus, the architect should create structures that collectively enhance the city's image and individually reflect the patron's status. To reach this goal, Alberti urges the practitioner to consider how buildings are perceived in their physical and social context.

Great individual buildings, like great individuals, attract attention. Alberti's own fame rests not only on his treatise but also on his association with important individuals and important buildings. Beginning in the 1450s, Alberti aligned himself with Giovanni Rucellai. A wealthy Florentine merchant and humanist concerned with learning and social status, Rucellai aptly fulfilled the author's definition of an ideal patron. He wrote, "I think I have given myself more honor, and my soul more satisfaction, by having spent money than by having earned it, above all with regard to the building I have done."[28] For Rucellai, Alberti designed the façade for Santa Maria Novella (1458–71) and the Palazzo Rucellai (ca. 1452?–70?), both in Florence (Figures 1.5, 1.6, and 1.8).

The Palazzo Rucellai in many ways reflects the ideas espoused by Alberti.[29] First, it was a good citizen. The magnificent new façade unified several adjacent, unmatched properties, imposing order where disorder had previously reigned. Second, the construct enhanced the patron's *virtu*. Standing in contrast to asymmetrical Trecento residences, the classicizing, harmonious facade reflected the power and erudition of Giovanni Rucellai, whose family symbol, the full-blown sails of fortune, blew across the front frieze. Third, the use of a trabeated system of ornament immediately identified the building as a residence of note.

The Palazzo Rucellai also reveals some drawbacks of the Albertian approach to architecture. Preoccupied with image, the architect and patron focused on the components with the greatest image-making potential. They were less concerned with utilitarian features, considering kitchens, plumbing, and storage only after the façade and interior public rooms had been designed. As a result, the structure's utilitarian components were left untouched when work halted. Behind the incomplete front elevation, the Palazzo Rucellai remains a hodgepodge of disorderly rooms and spaces.

Alberti argues that the practitioner should be praised for the ideas behind a building, not for the completed (or uncompleted) work. This focus predisposed the author to deal with process, not results. For all his careful analysis, Alberti does not consider that his experiments might have outcomes significantly different from those he desired. In effect, the learned author displays no interest in the predictive side of science. For example, he does not attempt to deduce the impact of existing social trends on architecture. Alberti acknowledges the breakdown of the extended family in *Della Famiglia* and calls for a return to traditional ways, yet he does not postulate what architectural changes would be necessary if family disintegration continued. Similarly, he calls for wealthy, learned patrons to make grand urban gestures, but does not consider the impact of such additions. Spurred by *De Re Aedificatoria*, competitive aristocratic patrons erected in-

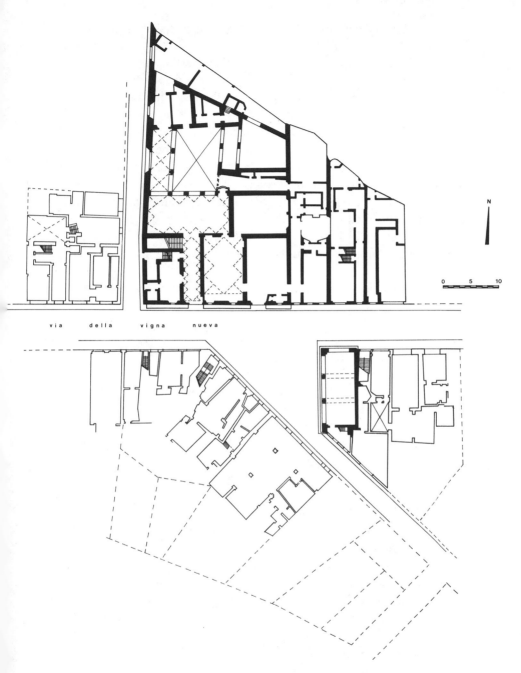

via della vigna nueva

N

0   5   10

Figure 1.8 Plan of the Palazzo Rucellai. (Mary Fishman.)

creasingly larger, grander palaces. The impressive, spacious salons of Quattrocento palaces expanded beyond any direct association with practical use. Single families rattled around in huge palazzi, vainly trying to meet high maintenance costs.

In considering the architects' people, Alberti set his calipers with great precision. Trained as a scholar, he was interested in the cerebral sides of architecture. The *praxis*, involving contact with lower-class people, did not appeal. Alberti calls for architecture to enhance the public good but in reality considers only the good of the elite. He portrays an architect of learning and refinement who interacts with other people of his own kind. Whereas later, more popularizing Renaissance treatises are written in the vernacular and filled with visually instructive images, Alberti's erudite treatise is in intellectual Latin and is scantily illustrated.[30] For him, architecture is "remote from the common use and knowledge of mankind" (VI.i).

The architects' people are a diverse lot. They include lowly workmen, uneducated users, wealthy clients, skilled engineers, style-oriented artists, and learned patrons. By focusing on a few of these participants, Vitruvius and Alberti reveal their philosophies and innate prejudices. Both operated within very class-conscious cultures. As a result, they interpreted all aspects of architecture and of the architects' people in relation to social ranking. In language, content, and form, the two authors specifically addressed the powerful aristocracy. They call for the architect to associate with and design for this erudite group. Only upper-class individuals have the knowledge to evaluate architectural achievement; only they have the money to fund grand projects; only they can acquire fame. Conversely, the two treatises argue that only a learned practitioner can satisfy the respected patron's desire for buildings with authority and magnificance; only he can write a respected treatise that will endure through time.

Readers of *De Architectura* and *De Re Aedificatoria* have no trouble identifying the most important of all the architects' people. The author stands supreme. The two texts are written in the first person; the authors' distinct personalities pervade every line. As a weary practitioner, Vitruvius longed for recognition. Educated, noble, but poor, Alberti desired the same. The two famous authors wrote to achieve personal *auctoritas* and *virtu*. To reach this goal, each consciously promoted the entire discipline of architecture, arguing for its categorization as an intellectual pursuit rather than a craft. By elevating the profession, they hoped to elevate themselves.

Besides the architect and patron, few other people inhabit the two treatises. In the stratified societies of Augustan Rome and Quattrocento Florence individuals reached upward to enter a higher social level; once there, they did not look down. Vitruvius and Alberti found it difficult to empathize with the lower classes. Lacking *auctoritas* and *virtu*, this segment of the architects' people was not worthy of discussion in learned texts. Furthermore, when the two authors wrote, tradition determined building forms; as a result, architects had no need to consult anonymous users. Similarly, practitioners had scant motivation for consorting with architectural laborers. Vitruvius and Alberti give the people who actually build the same perfunctory attention as a plumb line or any other tool.

With flourishes of erudition, Alberti and Vitruvius drew on the anthropocentric writings of the Greeks. In his treatise on painting, the Renaissance author acknowledges the famous precept of the Sophists, "Since man is the thing best known to man, perhaps Protagoras, by saying that man is the mode and measure of all things, meant that all the accidents of things are known through comparison to the accidents of man" (1.p.55). For Alberti, as for Vitruvius, the actions of man are the basis for architectural inquiry. The two authors maintain this focus, examining the specimens they know best; both measure the world and its inhabitants by measuring themselves.

Notes

1. Plato *Theaetetus* 160d. This and all other references to ancient texts are taken from the Loeb Classical Library unless otherwise noted.
2. See Frank Brown, "Vitruvius and the Liberal Art of Architecture," *Bucknell Review* 11 (1963):99–107; "Vitruvius," *Macmillan Encyclopedia of Architects*, Vol. 4 (New York: The Free Press, 1982), pp. 334–42. The best translation of *De Architectura* is by Morris Hicky Morgan, *Vitruvius: The Ten Books on Architecture* (New York: Dover Publications, Inc., 1960). All quotations are from Morgan unless otherwise stated.
3. Banned by tradition from putting their names on a temple they designed in Rome, two Greek architects found an ingenious solution. Sauras (lizard) and Batrachus (frog) carved their namesake creatures on the temple's frieze (Pliny *Natural History* 36. 42).
4. There is no evidence Vitruvius received any large-scale imperial commissions as a result of his writing. His only known architectural project is the municipal basilica at Fano described in his treatise (5.1.6) (Figure 1.2).
5. On the educational system outlined by Vitruvius see Brown, *Bucknell Review*, 99; R. L. Scranton, "Vitruvius's Arts of Architecture," *Hesperia* 43 (1974):494–99.
6. On optical theories in antiquity see J. J. Coulton, *Ancient Greek Architects at Work* (Ithaca, N.Y.: Cornell University Press, 1977), pp. 108–12.
7. Only later in the Empire did great pragmatic projects, such as aqueducts, come to be associated with *auctoritas* (cf. Front, *Aq.* 1.16).
8. The total number of properties acquired in Cicero's lifetime was probably much higher; see Israel Shatzman, *Senatorial Wealth and Roman Politics* (Brussels: Latomus, 1975), pp. 404–25. Cicero discusses various properties in letters to his friend Atticus (cf. *ad Atticus* 14.13; 15.17). Writing to his brother Quintus, he describes the state of the villa construction undertaken by Diphilus (*ad Quintum* 111.1.2).
9. On medieval interpretations of Vitruvius see Kenneth J. Conant, "The After-life of Vi-

truvius in the Middle Ages," *Journal of the Society of Architectural Historians* 27 (March 1968):33–38; H. Koch, *Vom Nachleben des Vitruvs* (Baden-Baden: Verlag für Kunst und Wissenschaft, 1951), *passim*.

10. On the life of Alberti consult the old but valuable work of Girolamo Mancini, *Vita di Leon Battista Alberti* (Rome: Bardi 1911 [1967]), and the entry by Eugene Johnson in the *Macmillan Encyclopedia of Architects*, Vol. I (New York: Macmillan, 1982), pp. 48–59. Unless otherwise noted, translations are taken from Leon Battista Alberti, *Ten Books on Architecture*, trans. James Leoni, ed. J. Ryckwert (London: Alec Tiranti, 1955).

11. Alberti wrote the *Descriptio urbis Romae* in the early 1430s, *Della Pittura* ca. 1436, *Della Statua* ca. 1435, and *Della Famiglia* ca. 1441.

12. On *magnificentia* see A. D. Fraser Jenkins, "Cosimo de' Medici's Patronage of Architecture and the Theory of Magnificence," *Journal of the Warburg and Courtauld Institutes* 33 (1970):162–70.

13. Giuseppe Martini, ed., *Momus o del Principe* (1443–52) (Bologna: Nicola Zanichelli, 1942), p. 150.

14. Author's translation of quote in William Harrison Woodward, *Studies in Education During the Age of the Renaissance 1400–1600* (Cambridge, England: Cambridge University Press, 1965), p. 49.

15. Anthony Blunt, *Artistic Theory in Italy 1450–1600* (London: Oxford University Press, 1962), p. 19.

16. Joan Gadol, *Leon Battista Alberti: Universal Man of the Early Renaissance* (Chicago: University of Chicago Press, 1969), pp. 130–32.

17. For his treatise on painting, Alberti studied Florentine artists at work, executed several paintings of his own, and experimented with perspective; see *Leon Battista Alberti on Painting*, trans. John Spencer (New Haven: Yale University Press, 1956), pp. 51, 55, 67, 70–71, 77–78.

18. The exact date of the apse is unknown but may be associated with Alberti's perspective studies for *Della Pittura*, completed a few years after he became prior of San Martino in 1432.

19. Although Vitruvius apparently did not gain significant commissions as a result of his treatise, Alberti did. In the 1450s he was involved in the reworking of the Borgo in Rome planned by Nicholas V; cf. Carroll William Westfall, *In This Most Perfect Paradise* (University Park: Pennsylvania State University Press, 1974). In the same period, he was commissioned to design the Tempio Malatestiano in Rimini and the Palazzo Rucellai in Florence.

20. Alberti mentions no living exemplars, not even his friend Brunelleschi; cf. Richard Krautheimer, "Alberti and Vitruvius," *Studies in Early Christian, Medieval, and Renaissance Art* (New York: New York University Press, 1969), pp. 42–52.

21. Quoted in Gadol, *Leon Battista Alberti*, p. 117.

22. Alberti provided a Tuscan translation of *Della Pittura* for his friend Brunelleschi soon after the Latin version was completed; Gadol, *Leon Battista Alberti*, pp. 215–16; John Onians, "Alberti and Filarete, A Study in their Sources," *Journal of the Warburg and Courtauld Institutes* 34 (1971):96.

23. For the debate over architectural expenditures in the early Renaissance see Fraser Jenkins, *Journal of the Warburg and Courtauld Institutes* (1970):162–70; Richard Goldthwaite, "The Florentine Palace as Domestic Architecture," *American Historical Review* 77 (October 1972):990–1012.

24. Leon Battista Alberti, *The Family in Renaissance Florence* (Columbia, SC: University of South Carolina Press, 1969), p. 210. See also *De Re Aedificatoria* (IX. i); Joseph Rykwert, "Inheritance or Tradition?" Profile 21: Leon Battista Alberti, *Architectural Design* 49 (1979):4–5; Onians, "Alberti and Filarete," 97–104.

25. Alberti relied heavily on overseers in his own work, as is evident in his correspondence with Matteo dei Pasti on San Francesco in Rimini; see Leopold Ettlinger, "The Emer-

gence of the Italian Architect During the Fifteenth Century," in *The Architect*, ed. Spiro Kostof (New York: Oxford University Press, 1977), pp. 113–14.

26. Surprisingly, Alberti does not discuss competitions, even though several major contests occurred during his life.

27. Alberti did allow for variations when necessity dictated. For example, he admitted the use of arcuation for grand public basilicas in order to span large areas (VII. xiv).

28. Quoted in Goldthwaite, "The Florentine Palace as Domestic Architecture," 990–91.

29. On the Palazzo Rucellai see Charles Mack, "The Rucellai Palace: Some New Proposals," *Art Bulletin* 56 (December 1974):516–29; Kurt Forster, "The Palazzo Rucellai and Questions of Typology in the Development of Renaissance Buildings," *Art Bulletin* 58 (March 1976):109–13. On Giovanni Rucellai see Richard Goldthwaite, *The Building of Renaissance Florence* (Baltimore: Johns Hopkins University Press, 1980), p. 88.

30. For example, Sebastiano Serlio (1475–1555) wrote the *L'architettura* in Tuscan and included ample illustrations for the illiterate. Although acknowledging contemporary hierarchies, he gives attention to lower-class buildings as well as to medieval and French variations; M. N. Rosenfeld, "Sebastiano Serlio," *Macmillan Encyclopedia of Architects*, Vol. 4, pp. 37–39.

2

Wright's Written People

RUSSELL ELLIS

Who were Frank Lloyd Wright's "people" in the sense of this book's theme? Two features of Wright make him an interesting prospect for scrutiny. First, he is America's most renowned architect: admired, read, and studied more deeply than any other in our history. Second, he made the investigation of his ideas on architecture easy. Starting in his unhappy mid-career, Wright embarked on a project of writing—and later, lecturing—of major proportions, modeled no doubt in part on the romantic wordification of architecture by his mentor Louis Sullivan.

One particular set of writings is especially interesting. In 1908, Wright wrote the first of a series of fourteen articles commissioned by the most important journal in his field, *Architectural Record*. Directed at a professional audience, these articles appeared between 1908 and 1952. Frederick Gutheim, editor of his collected writings and of the assembled volume of these articles, contends, "it is in these writings alone that Wright has made a deliberate effort to formulate the principles of his art (Gutheim, 1975, p. VIII). I have examined *In the Cause of Architecture* to discern the place of people in the realm of his architectural teachings.

Wright's fortunes and architecture changed over the years, but there is a remarkable consistency in the style and content of his statements. Indeed, he is oppressively repetitious and appears to have said all that he had to say, thematically, on his topic. Adapting Herbert Muschamp's stance to my purposes, I have conducted a content analysis of this exhaustive collection:

The phenomenon of an architect's public persona, the relation of his philosophical intentions to the built results, his use of the tools of mass communication to project a private vision, the question of his social motivations, are all speculative areas of limited value in objectively assessing the formal achievement of an individual building. At the same time, as indicators of the way an architect con-

ceives and structures his relations to society, these issues are highly pertinent.
. . . (Muschamp, 1983, p. 18)

Wright's Written World

Entering Wright's world of thought, one encounters poetry effortfully echoing the stentorian and magisterial voice of Walt Whitman, whom he admired, and the mystic philosophising of Giorgi Gurdjieff, to whom he was exposed through his wife Olgivanna (Twombly, 1973, pp. 147, 172, 176). Interlarded throughout is a contentious refrain of historic wrongs and contemporary—mainly Wrightean—rights. In a celebration of modern American democracy, human nature, flowers, trees, social order, materials, and machines gather, under the imaginative guidance of the "creative artist," into a surging "organic" architecture that, according to Wright, not only reflects modern life but will save it.

> All man has above the brute, worth having, is his because of Imagination. Imagination made the Gods—all of them he knows—it is the Divine in him and differentiates him from a mere reasoning animal into a God himself. A creative being is a God. There will never be too many Gods.
>
> Reason and Will have been exalted by Philosophy and Science. Let us now do homage to Imagination.
>
> Imagination is so intimately related to sentient perception—we cannot separate the two.
>
> Let us call Creative-Imagination the Man-light in Mankind to distinguish it from the intellectual brilliance. It is strongest in the creative-artist. A sentient quality, and to the extent that it takes concrete form in the human life, it makes the fabrication live as a reflection of that Life any true Man loves as such—Spirit materialized. (Gutheim, 1975, p. 145)

Wright's people take shape, by indirection, against the backdrop of wood, stone, steel, and glass wrought by force of the new democracy into a "foil for life." What is the human nature best expressed and animated by it?

Human Nature and Purpose

Like Marx, Wright viewed human nature as historically mutable. Conditions set by the new democratic society allow the emergence in life and built form of long-buried human potential.

> Materials everywhere are most valuable for what they are—in themselves—no one wants to change their nature or try to make them like something else.
>
> Men likewise—for the same reason: a reason everywhere working in everything.
>
> So this new world is no longer a matter of seeming but of *being*.

Where then are we? . . . [In] the first Democracy of *being* not seeming.

The highest form of Aristocracy be it said the world has ever seen is this Democracy, for it is based upon the qualities that make the man a man. (Guthein, 1975, p. 145)

"Men likewise. . . ." Wright's people seldom appear to us except as mediated through a material or vegetal equation. The emergence of man, material, nature, and civilization is always portrayed as an organic whole; as aspects of each other. Occasionally, humans possess autonomous features. Certain persistent qualities in human nature are freed in "this new world . . . of *being*."

Human feeling loves the vigor of spontaneity, freshness, and the charm of the unexpected. In other words, it loves life and dreads death. (Gutheim, 1975, p. 135)

Also, through the arts and creative artists (especially architects) we are now in a position to realize our love of life in our poetic nature.

Any of these Arts called "Fine" are Poetic by nature. And to be poetic, truly, does not mean to escape from life but does mean *life raised to intense significance and higher power.*

Poetry, therefore, is the great potential need of human kind.

We hunger for Poetry naturally as we do for sunlight, fresh air and fruits, if we are normal human beings. (Gutheim, 1975, p. 225)

The purpose of our nature—our imagination, our love of life, our natural hunger for the poetic, and so on—is, Godlike, to realize life through manifestation of "the spirit": a goal common to all nature.

The quality of *life* in man-made "things" is as it is in trees and plants and animals, and the secret of character in them which is again "style" is the same. It is a materialization of spirit. (Gutheim, 1975, p. 133)

From this historically conditioned and exalted nature, the goal of life and living emerges.

Earth-dwellers that we are, we are become now sentient to the truth that living on Earth is a materialization of Spirit instead of trying to make our dwelling here a spiritualization of matter. (Gutheim, 1975, p. 151)

The key to Wright's people is understanding that the artist is the clearest distillation of the best human qualities. Creative imagination is the most developed in him. He is the true conduit through which spirit is materialized. Achievement of the core purpose of our dwelling on earth is his calling. This is important to keep in mind when we try to envision Wright's homunculus occupying built things.

Social Order

Wright is a polemical theorist of social order. He is a celebrant of democracy in the United States ("Usonian Democracy") and a severe critic of slavish invocations of antiquity, in architecture and society. They are enemies of creative imagination and the new life.

American democracy is liberation of the spirit. Its unique structure opens dramatic new possibilities for individuality, expression, and architecture. It is the true historic setting for the emergence of human potential. The old is death.

> I do not believe we will ever again have the uniformity of type which has characterized the so-called great "styles." Conditions have changed; our ideal is Democracy, the highest possible expression of the individual as a unit not inconsistent with a harmonious whole. The average of human intelligence rises steadily, and as the unit grows more and more to be trusted we will have an architecture with richer variety in unity than has ever arisen before; but the forms must be born out of our changed conditions, they must be *true* forms, otherwise the best that tradition has to offer is only an inglorious masquerade, devoid of vital significance or true spiritual value. . . . (Gutheim, 1975, 56)

But Wright takes a peculiar turn. Individuality in a harmonious whole is a delicate point. Indeed, it is the crucial issue in any sociopolitical vindication of democracy in a mass society. But the working dynamic in this democracy is, for Wright, the very fact of standardization and its symbol, the machine.

> Standardization as a principle is at work in all things with greater activity than ever before.
>   It is the most basic element in civilization. To a degree it is civilization itself.
>   *Standardization* should have the same place in the fabric we are weaving which we call civilization—as it has in that more simple fabrication of the carpet. (Gutheim, 1975, p. 135)

> The machine is the normal tool of our civilization, give it work that it can do well—nothing is of greater importance. (Gutheim, 1975, p. 55)

Given the romantic centrality of poetry, imagination, and hunger for spontaneity and freshness that vivify creative human nature, one would assume the leveling and cheapening tendencies of standardization to be a peril. Indeed, Wright warns vigorously of this prospect, but creative imagination freed by the new democracy renders standardization a positive and *organic* element in the work of the architect!

> Standardization can be murderer or beneficent factor as the life in the thing standardized is kept by imagination or destroyed by the lack of it . . .

The "*life*" in the thing is that quality of it or in it which makes it perfectly *natural*—of course that means organic. And that simply means true to what made it, as it was made, and for what it was made. That would be the *body* of the "thing." (Gutheim, 1975, p. 136)

Social Order in Architecture

Wright is effusive and certain about the implications of this social world for the array, structure, and function of the new architecture. Despite his broad social analysis, however, we know only by the *fiat* of poetic juxtaposition that their connections exist. The how and why remain unexplicated. Usonian democracy stretches and shapes plans and forms into mutually constant shapes; into organic architecture. But the profoundly influential *patterning* of life is so distantly implied as to be indistinguishable from the architecture itself. Certainly, democracy, standardization, and the machine are, in his rolling logic, something like causal, but buildings appear, in his vision, to be the active agents.

> Shimmering, iridescent cages of steel and copper and glass in which the principle of standardization becomes exquisite in all variety.
> Homes? Growing from their site in native materials, no more 'deciduous' than the native rock ledges of the hills, or the fir trees rooted in the ground, all taking on the character of the individual in perpetual bewildering variety.
> The City? Gone to the surrounding country. (Gutheim, 1975, p. 149)

The American Middle West, "Cradle of Democracy," gave birth to Wright's organic architecture. But the historic origins of his architecture do not hint at the specific social processes served, although its entire purpose is to serve them.

> Gradually, over a fifty-year period . . . it planted and established fertile forms and new appropriate methods for the natural (machine) use of steel, glass, plastics (like concrete) and provided more ample freedom in shelter for the free new life of these United States than any "style" had ever provided or even promised. Organic-architecture thus came of America—a new freedom for mixed people living a new freedom under a democratic form of life. . . . Organic-architecture was definitely a new sense of shelter for *humane* life. (Gutheim, 1975, p. 233)

This broad conception linking an emergent society and architecture shrouds any formative dynamic we might seek. "Fertile forms" and "appropriate methods" are "planted," creating "a new freedom for a mixed people." A new architecture results: "Shelter for humane life." One need not be critical of the historical adequacy of a confident architect. The point here, in relation to our quest, is that we do not find collective *homunculi* in Wright's history.

Interestingly, organic architecture does somehow reshape buildings to fit a heterogeneous democratic people. We are left to guess at the daily imperatives of their lives to imagine the necessary redesign of their shelter.

> Even the walls played a new role or disappeared. Basements and attics disappeared altogether. A new sense of space in appropriate human scale pervaded not only the structure but the life itself lived in it was broadened, made more free because of sympathetic freedom of plan and structure. The interior space to be lived in became the reality of *the whole performance*. (Gutheim, 1975, p. 234)

The image here is enticing but not suggestive: a broad and free life facilitated by plan and structure "sympathetic" to that life. In our search for the social life housed in this new society, we must accumulate features of the "house" itself.

> Buildings perform their highest function in relation to human life within and the natural efflorescence without; and to develop and maintain the harmony of a true chord between them, making of the building in this sense a sure foil for life, broad simple surfaces and highly conventionalized forms are inevitable (Gutheim, 1975, p. 61)

Something in the new life makes broad, simple surfaces and highly conventionalized forms inevitable. What causes these forms is not clear. As far as we can tell from the text, it is the inexorable march of standardization in some complex interplay with the social dynamics of democracy.

Buildings and People

Only when we ask about the dwelling do we find instructions; vague with respect to the interactive nature of occupants, but definite in its directives to architects. One overarching principle is consistent with Wright's conception of American democracy:

> There should be as many kinds (styles) of houses as there are kinds (styles) of people and as many differentiations as there are different individuals. A man who has individuality (and what man lacks it?) has a right to its expression in his own environment. (Gutheim, 1975, p. 55)

However, a more specific programmatic directive, not discernibly related to any other point in the text, tells us:

> The most truly satisfactory apartments are those in which most or all of the furniture is built in as a part of the original scheme considering the whole as an integral unit (Gutheim, 1975, p. 55)

The sole remaining *dictum*—and perhaps Wright's most influential—articulates the dwelling plan with great concreteness.

A building should contain as few rooms as will meet the conditions which give it rise and under which we live, and which the architect should strive continually to simplify; then the ensemble of rooms should be carefully considered that comfort and utility go hand in hand with beauty. Beside the entry and necessary work rooms there need be but three rooms on the ground floor of any house, living room, dining room and kitchen, with the possible addition of a "social office"; really there need be but one room, the living room with requirements otherwise sequestered from it or screened within it by means of architectural contrivances. (Gutheim, 1975, p. 54)

A house for Usonian people. Let us now try to give these people the features, needs, and impulses implied in what Wright has led us to understand. How can we blend the house into their work?

Wright's Homunculus

Attempting to discover Wright's people in these writings, we can retrace our steps, examine the terms he uses to describe them, and relate the distilled image to what we can make of organic architecture. Listed here, in the order they were encountered, are these descriptive terms:

Gods	pure
poetic	gracious
joyous	dignified
genuine	sincere
natural	lovable
free	true
integral	like trees
spiritual	like flowers
individual	

Repeatedly, these are the words Wright uses to describe people *and* "the best" architecture. Buildings and people at their best are constantly equated: "A building has a presence as has a person . . .", "Buildings like people must first be sincere, must be true and then withal as gracious and lovable as may be." People and buildings are also equated with trees, flowers, and, on occasion, elements in a graced landscape. Human character and the character of their dwellings are organic and spiritual facts whose natural *raison d'être* is the materialization of "spirit." All is participation in God's work here on earth. Indeed, we are gods.

The core bearer of all these qualities, summarized in "creative-imagination," is the artist (architect). They are "strongest in the creative-artist (architect)." Our material fabrications achieve their spirituality through this medium, but "[to] a degree all developed individuals have this quality." Our development of these qualities grows from the social soil of democracy.

With few exceptions[1] Wright's individuals are classless, "average Americans," freed, in the new democracy, of the need for radically differentiated dwellings. More open to each other; to direct, unmediated contact. But we can't discover why.

Embedded in a democracy of trustworthy individualism, and having shed the classical "grandomania" of the aristocratic, monarchic, and oligarchic past, this new being is in need of human-sized buildings.

> In the matter of scale, the human being is the logical norm because buildings are to be humanly inhabited and should be related to human proportions not only comfortably but agreeably. Human beings should look as well in the buildings or of it as flowers do.
>
>   People should belong to the building just as it should belong to them. (Gutheim, 1975, p. 154)

Otherwise, this being is ageless, sexless, raceless, and groupless and has no personality save the poetic grandeur, dignity, and so on, potentially shared by all the new society, or features discovered by the architect in the great variety of individuals this society permits.

Indeed, Wright's *homunculus* is, in part, the society; in part, the *living* building; in part, an incomplete approximation of the architect. All is woven into our fabrications. The *homunculus* never truly walks or sits in them but stands majestically with them mutually fused in individual dignity, spiritual and organic beauty. It is a heroic pose.

> As though a wave of creative-impulse had seized stone and, mutable as the sea, it had heaved and swelled and broken into lines of surge, peaks of foam—human-symbols, images of organic life caught and held in its cosmic urge—a splendid song! (Gutheim, 1975, p. 175)

One structure clearly meets the qualities identified in these essays. Some single Wright building may, but it is Richard W. Book's sculpture for the hallway of the Dana House "Flower in the Crannied Wall" (Gutheim, 1975, p. 119) that fully comprehends Wright's written vision. Here a slender pyramid, decoratively faithful to Wright, rises to spires lovingly hovered over by a graceful feminine figure, her delicate hands apparently adding the last fine element of a spire. She and the building are fused from midriff down.

Conclusion

It will be clear to any reader of the *Record* essays that Wright was free-ranging across what he conceived as the fundaments of architecture in an evolving world. People do not figure in or around his ideals of architectural

Figure 2.1 A near-perfect representation of the romantic and heroic relationship between Wright's people and buildings is Richard Bock's sculpture "Flower in the Crannied Wall," executed under Wright's direction for the Dana House hallway. (Reprinted from *In the Cause of Architecture,* an Architectural Record Book, © 1975.)

form. These forms *are* animated but with ideal features of humanity and earth brought by the bearer of the best human qualities: the artist/architect. Governments, class, eras are anonymously good or bad in their indistinct impact on the "best" in buildings.

What then is to be made of this investigation? Robert Gutman carefully instructs us in the dangers of interpreting architects' words (Chapter 5). And Edgar Kaufman, Jr., responding to an earlier version of this study, noted, "It did seem to assume that architects are more interesting when working on projects than on real commissions, or when theorizing than when describing actual experiences."[2] One is chastened by such observations.

Architects are certainly not more interesting, when examined in this light, but they are distinctively interesting in light of contemporary social questions applied to their thought and works. Seldom—and certainly not in Wright—will we find the isomorphy of social thought and architectural intentions Bloomer uncovers in Kerr's Bearwood (Chapter 6).

But two interesting research questions remain. First, would more be revealed about Wright's conception of people through a detailed investigation of specific commissions, investigations such as that carried out on Kahn's Richards Laboratory by Robert Gutman (Chapter 5)? Second, assuming the texts I have examined here are typical, how did these indistinct conceptions of people figure in the conceptions of Wright's array of followers? Wright offers tantalizing opportunities for interpretive license: "A building should contain as few rooms as will meet the conditions which give it rise"; "a new sense of shelter for *humane* life"; "a new freedom for a mixed people. . . ."

Finally, one must entertain the possibility that only a study of the social assumptions of the times contained in Wright's eclectic collection can reveal an image of life conduct, or the possibility that Wright had no images.

Notes

1. Wright makes very few social distinctions throughout the text. The most significant are identified in the following:

> The average American man or woman who wants to build a house wants something different—"something different" is what they say they want, and most of them want it in a hurry. . . . The average man of business in America has truer intuition, and so a more nearly just estimate of artistic values, when he has a chance to judge between good and bad, than a man of similar class in any other country. But he is prone to take that "something different" anyhow; if not good, then bad. He is rapidly outgrowing the provincialism that needs a foreign-made label upon art. . . . (Gutheim, 1975, p. 125)

2. Letter to R. Ellis, August 26, 1983.

References

Gutheim, Frederick, ed. *In the Cause of Architecture*. New York: Architectural Record Books, 1975.

Muschamp, Herbert. *Man About Town: Frank Lloyd Wright in New York*. Cambridge: MIT Press, 1983.

Twombly, Robert C. *Frank Lloyd Wright: An Interpretive Biography*. New York: Harper & Row, 1973.

Joseph Esherick:
The Drama of the Everyday

DANA CUFF and RUSSELL ELLIS

As a kid, I used to get up at five Saturday mornings, chase the neighborhood cats for awhile, and then take the bus [from Philadelphia] into New Jersey where there were a bunch of old farms. I would wander all over the place doing drawings of buildings. I assumed they were buildings *for* somebody, and that somebody had a very attractive idea (to me) of how to live. They had porches, generally for shade, they had gardens and the like. Generally speaking, since these houses were built by the people who were going to live there, they really paid attention to making it comfortable and making it a pleasant place. There was a certain amount of stuff for show, but for the most part, they were people-oriented.

The magnet of those houses for me was the idea of living in them. Not necessarily me, just someone. Well, I suppose I thought that I might be living in the place. As a kid, I'd look at these old farmhouse wrecks, and then do measured (put that in quotes) drawings (also in quotes) and then I'd reconstruct my conception of how someone would live there. Most of these houses were abandoned, so I could get inside to see how they were built and all the tool marks. They were usually pre-plumbing wrecks, so then I'd figure out where the kitchen and bath would go, and how the roof would have been, where the entry should be and all that sort of thing. They were fascinating because old buildings have a kind of soft, used quality about them that is very hard to duplicate in a new building. And the other thing was that they were *there*. It was a lot easier for me to start with something there than to create something purely out of my imagination.

Joseph Esherick, 1989 recipient of the A.I.A.'s Gold Medal, founding partner of the San Francisco firm Esherick, Homsey, Dodge, and Davis, which won the 1986 American Institute of Architecture's Firm of the Year award, speaks of the evolution of his ideas about architecture and about the life it houses.[1] Joe's views about the way buildings are put together, about the design process, about materials, and about the ordinary have been written about before.[2] Here we asked Joe to tell us about his notions of people, to

55

Figure 3.1 Entrance to The Cannery, 1966, San Francisco. (Esherick Homsey Dodge and Davis.)

provide another perspective on some of those same issues. For example, the hearty, practical folk who built and inhabited his farmhouses capture the same quality of a "special ordinariness" that Charles Moore saw in The Cannery, when he likened it to a tea ceremony (see Figures 3.1 and 3.2).[3]

Our report of this interview, although not always following the free-floating conversation itself, begins with Esherick's description of his imaginary people's origins. From these roots, we see that he places a simultaneous emphasis on the relationship between the unique individual and the generalized standard or society at large, and the ordinary. As an architect, he takes the individual, the norm, and the ordinary into account by being a careful observer of the world around him. He has developed a working approach to portray future settings in terms of the immediately visible. In this way, current clients and his own everyday experiences serve as resources for peopling buildings that have unspecified inhabitants. Esherick understands the relationship between people and things, between people and buildings, to be an organic one that can be best described as drama.

Figure 3.2 Multileveled activity at The Cannery in San Francisco. (Esherick Homsey Dodge and Davis.)

He concludes by suggesting how we could train future architects to develop a vital set of mental actors. Thus, there are five interrelated themes that emerge from our conversation with Joe Esherick: the formative influence of personal biography, the relationship between the unique individual and the norm, the significance of the ordinary, the relationship between the seen and the unseen, and drama as a link between people and things.

"I think it was significant that [my uncle] Wharton insisted on certain things, like chairs being comfortable, and the fact that everybody wasn't the same size. He designed couches with a narrow end and a deep end, so you selected a place either by how you felt or by your physical features." Esherick credits Wharton Esherick, the sculptor, with many insights not only about people and comfort, but about materials and craft. Literature also had a significant effect on the evolution of his point of view. "The influence of literature on me is considerable. Especially the novels of particular authors like E. M. Forster, James Joyce, or W. H. Hudson and Henry Handel Richardson (who was a woman, you know). All these people fascinate me with the way they think, and with the way they construct people and places. Sometimes I think places are made better in literature than in architecture."[4]

From the authors and Wharton, Esherick learned to view people as particular individuals.

> I have some ideas about how you really get to know people, clients. I used to devise questionnaires to give clients, but basically what I got back were the questionnaires filled out! Responses. The questions had all been filtered through me and the problem is to figure out how to filter the questions through the client. Now I go with clients to look at their lot—it's a big arm-waving exercise, but I get to understand that they like the view down the valley, even though I may think the one up into the mountains is best. I can comment that I like that view, in case they haven't noticed it, but the view I like doesn't really have anything to do with it since they're the ones that are going to be looking at it. At the site I can also provide technical assistance, about wind and things like that. Then I go to the drawing board and I start with a survey and a blank piece of paper, talking about what they want. And I do the drawing there, with them. It's very scary because sometimes I don't have any ideas. But I can tell instantaneously whether they understand what I draw and like it, or whether it makes them uneasy. I am getting to know them, not in a general sense but in terms of whether they like the idea of a ten-foot ceiling or an eight-foot ceiling. Everyone's idea of beauty, I think, ought to be assumed to be different and individual—that's my assumption anyway. I would like my buildings to make what it is the people want to do easier to do—more pleasant, more lively, more beautiful. Fortunately, we don't work for many people who want to make life uglier. I think buildings should help people to operate on their own and individually, rather than to force them into molds or cliques and so on.

Esherick elaborated on this notion of the individual in relation to the common elements of society in a discussion of some apartments he had recently completed.

> There are things like pollens drifting through the whole society, and everyone sneezes at once (over white walls, or Thai silk). Some of these are ephemeral, and others are more permanent (or have always been there: that you like it to be warm). I'm informed by that, and I lay it on these poor bastards who've yet to come [the prospective apartment tenants]. I've been as attentive as one can be about why people rent apartments. Not everybody wants a view. There are these dumbell plans in San Francisco, and in the center are quiet little apartments that give onto the light court; they don't look at anything. Those apartments rent to people who want some quiet and privacy, who don't care about a view. They want this little internal shell. Conceivably they're people who are bartenders who work all night, and who want quiet, dark places so they can sleep in the daytime. It's clear to me that you don't design space for everybody, and there's a relationship between what some people are going to gravitate toward and the particular design, which relates to the place, its orientation, its acoustic environment, to a whole lot of things. You can enhance the situation by exaggerating the particular

conditions and magnetizing people selectively. For instance, you don't paint Niagara Falls on the wall across from the apartment in the light court, because that person doesn't care about the view. So you make all your design decisions more consistent with that particular point of view, as long as you have a conception that there is somebody (and enough of those people) who will appreciate this [exaggerated, nonstandardized scene].

The real problem is not to just make things livable, for God's sake (that always makes me think of cutting off the oxygen supply). Without sounding pompous, the architect's main function is to bring beauty to the thing. Not a personal vision of beauty, but a beauty that really works, that makes the people who see it feel better and enjoy what they're doing without being dominated by the damn thing. If you want to be dominated by something you go to a museum, but you don't want to live in a museum.

Esherick here defines beauty in human terms. The sense is that the ongoing, everyday life is heightened by beautiful settings.

It isn't a question of elevating things. I don't agree with [San Francisco architecture critic] Temko's argument about "noble materials." What I try to do is to get people to see the potential beauty in anything, like the beauty of a worn concrete sidewalk or an asphalt street. Concrete or asphalt is not automatically bad; it is a question of how you use those things. We need to give ordinary, common things their place because, you know, the bricks *are* holding up the wall. You don't try to make more of them than what they are, you try and let them be what they are. I've always believed in doing things in very ordinary materials, in what's available, because of economy and familiarity. If it's unfamiliar material, people won't understand what their relationship to it ought to be. Chris Alexander has this idea that windows ought to be cut up into small panes. I'm not sure what his argument about it is, but my argument is that you don't necessarily cut up the window into small panes. But if you look out a window and you understand that it is a window when you're looking out of it, then you look out of your window and you see that there are other windows out there that look like the one you're looking through. I believe this *event* attaches you to whatever is out there—which I regard as a good thing—so I like to do windows that are like everybody else's windows.

To arrive at some understanding of the everyday and ordinary, Esherick implies that one must go out and look at the world.

One good thing about the Beaux Arts, at least at Penn: they emphasized that you go out and constantly observe what people were doing, and look at places and talk to people there. Teachers like James House, John Harbeson, and Paul Cret himself were oriented this way. I remember going out to the huge, monumental flight of stairs at the Philadelphia Art Museum. The stairs are so wide they're ridiculous; you can march the tenth infantry division down the stairs and it would probably only take about three rows of them. They're just tremendous things. But you look at these steps, and watch people go up them: they may start up straight, but they actually go up in a kind of parabola. And everybody does the same thing—as soon as they start to get tired, they adjust their course to what's

Figure 3.3 The shop window at Very Very Terry Jerry. (Photograph © 1967 Jeremiah O. Bragstad, San Francisco.)

comfortable. It was really dramatic to see, something I still like to watch. It's a good check on whether stairs are right or not.

Like the tenants, the stair climbers are an example of the relationship between the observable and the unknown. By attending to the ordinary aspects of human activity in everyday settings, the architect collects a resource of useful observations. "The images of the unidentified person—someone that you don't know who is going to occupy a building—come from two sources: one is general observation and the other is the very intimate detailed relationships with specific clients on their houses. This goes back to the assumption about pollens in society."

The frame that Esherick often places these observations within is a kind of theatrical metaphor.[5] He describes this view as he talks about one project, a women's clothing store called Very Very Terry Jerry (Figure 3.3).

The clients were a really stimulating pair—one was a song and dance guy who had a long experience on Broadway and the other had grown up in New York's garment industry. They didn't talk about what the place should look like, but about how they sold clothes. They had an idea: to use men to sell women's clothing, with the theory that the man could flatter the woman more and that a relationship could build up. This was the 60s, and I thought these guys were really on to something. I saw it really as theater. It was almost an exaggeration of a Goffman dramaturgic hypothesis. So I sat in their New York store, as if I were a husband waiting for his wife, and watched this whole scene. A woman would come into this store in her own clothes, and put on something else, and with the help of these sales guys, would be completely transformed. She struts up and down, turns around, and looks at herself in the mirror. She's *really* on stage, and goaded on by the Brechtian-like commentator sitting at the side. The usual way was to do this in a private setting. What we worked out was not to have any clothes in the store show window—nothing but the mirrors. So this act is what happens in the show window. All the active stuff was out on the outside, where people coming along could see it. The women didn't object since the style of the clothes was very flamboyant, so I was trying to match that. Even the place where the make-up expert transformed faces—we put that in the window too. It worked wonderfully well. It really was a great show. People would gather outside watching all this crazy stuff. I don't think anyone has ever done this before or since.

I had been interested in this sort of thing before because I got interested in restaurants and had been watching waiters. I used to station myself in the kitchen, where you see the waiter out putting on his dining room act, and the minute he comes through the doors—he's completely transformed. His role is to fight with the cooks. He takes the napkin that has been on his arm and starts snapping it around. The smile is gone and he's yelling and screaming. A total transformation. If he came into the kitchen smiling, the cooks wouldn't know what to do with him. But this dramatic role-taking was interesting. My interest in looking at the waiters was to create a scene where the waiter could carry out his role in a way that really interested him, without irritations, because then he would provide the best service possible and you establish the best possible relation for the client. You can't just focus on the customer and forget the waiter.

As such, from the clothing store to the restaurant, we see that actors, stage, props, and audience are all part of an interconnected, organic system. Roles are taken, not automatically, but with some encouragement, and architects can compose the setting in such a way as to promote a vital set of roles and interactions.

Indeed, the role of the designer appears to be to establish a relationship between people and things that is respectful of their ongoing situation, that heightens experience, that creates from what already exists. Esherick talks about these views in terms of the future of design education.[6]

We should teach how to observe people, how to talk, but more important, how to really listen to people and how to get at people's feelings about things. There may be some stable material you could teach, probably physical things (you can only stand so much glare; if your living room gets below 40 degrees you're going to be cold . . .). But not those damn studies that show where people sit in libraries and how they behave in elevators. We should put those in a zoo.

People should understand about the whole issue of contamination, since most experiments are contaminated. The question itself is the original contaminant. We have to learn to remove as much as possible. And someone needs to talk about the ethics of intervention: how to be honest and useful and avoid being autocratic and possessive. In school, we encourage a defense–attack relationship with the review system, where a bunch of professors sit with their backs to the students. It's like any hazing or rite of passage, and it gets carried on into architectural practice.

All of this must be woven into the architect's primary function, which is not to stand around making observations and collecting data, but to *use* it. We need quick and dirty methods, or else we need to inculcate students with the notion of continuous responsibility. Like the bus ride I took this morning to the office. I was watching people choose seats, and I enjoyed trying to figure it out. I didn't come to any conclusions, but that's not the point.

There's one additional set of ideas that seems to have some potential. A while ago, I read Umberto Eco's "The Role of the Reader." Some of it is pretty murky stuff, but he makes the interesting argument that after a book is written it passes out of the hands of the writer and into the hands of the reader. The reader then takes the book over and what he does in interpreting it is an extension of the author's activity. This seems to be a good way to express what I think is important: that an architect's work isn't either something he or she, the architect, owns, nor is it a fixed, immutable thing in the landscape—it is rather something that changes, or at least, ought to change. Certainly the interpretation and probably the mode of use will change. I'm uncomfortable with trying to do things that can only be interpreted in one way, or things that try to be instructive or influential. Such works end up being authoritarian, which I simply don't like. There is, perhaps, a tendency to believe that there is something permanent in the perception of great monuments like the Parthenon or Gothic cathedrals or whatever (or even the U.S. Constitution), but I don't think it works that way and I think it is futile to expect such permanence. It is probably worse than that—it suggests that there is no change, decay, or growth—that everything is written or cast in some kind of wear-resistant form.

There is a rich series of images in this conversation. Joseph Esherick's people, as he describes them, are complemented and enlivened by their settings. They are farm families, waiters, bartenders, shoppers, clothing salespeople, renters, bus riders, stair climbers. These people are mentioned along with certain physical entities like houses, bricks, and typical windows. What all of these people and things have in common is that they are ordinary. One is reminded of Wharton Esherick's chair made of ax handles, or of the coat pegs he carved into likenesses of the carpenters building his studio. Joe Esherick's challenge is to observe them—both people and things—well enough to "give them their place." Their very ordinariness places them in the common domain, so that they allow the architect to learn generalizable notions. It is through the specific and the ordinary—through stair-climbing patterns, bus seat selection, and the front–back regions of waiters—that Esherick connects that which he can observe with that which will be. New settings rest on the observed life of the old, observations that are elaborated and projected into the possible. "I would like my buildings to make what the people want to do easier—more pleasant, more lively, more beautiful." This is all part of an organic whole, a Goffman-type stage. This metaphor of situated life captures the interaction of people and things for Esherick. The architect is part of the drama, because that observation and filtering leads to action—a new stage.

Notes

1. Two articles review Esherick's office work in relation to the Firm of the Year award: Michael J. Crosbie, "Firm of the Year: Esherick, Homsey, Dodge and Davis." *Architecture* 75 (February 1986):28–37; and *Architecture California* 8 (July/August 1986):22–31.

2. See, for example, "Joseph Esherick: Theory and Practice," *Western Architect and Engineer* 222 (December 1961):20–27; and more recently, Esherick's chapter about his own training: "Architectural Education in the Thirties and Seventies: A Personal View," in Spiro Kostof, *The Architect* (New York: Oxford University Press, 1977), pp. 238–79.

3. Charles Moore, "The Cannery: New Old Marketplace in the City; How It Looks to a Critic," *Architectural Forum* (June 1968):76–79.

4. Raymond Lifchez makes the same point about places in literature in his chapter in this volume.

5. As Esherick states, Erving Goffman's dramaturgic metaphor is his model. See Goffman's *Presentation of Self in Everyday Life* (New York: Doubleday Anchor, 1959).

6. Esherick, as a member and past chair of the University of California, Berkeley, Department of Architecture, is known for his ideas on design education. He won the profession's annual award for excellence in education in 1982. See "AIA ACSA Select Esherick for Education Excellence Award," *AIA Journal* 71 (March 1982):35; and Esherick's chapter in *The Architect*, cited in note 2.

Through the Looking Glass:
Seven New York Architects
and Their People

DANA CUFF

A recurrent feud in architecture pits art and social commitment against one another. In the first camp, proponents argue the preeminence of expression, creativity, and the individual maker. Their opponents, comprised of critics, social scientists, and those claiming to represent the public interest, expound architecture's social responsibility. Between the two groups there seems to be little common ground. This study suggests, however, that the two positions are united in a fundamental dialectic, bound by the reciprocity that exists between an individual's identity and the individual's sense of others. The image of the individual and the image of society, or the self and the other, are mirror reflections. Social psychologist George Herbert Mead and later, Alfred Schutz, a sociologist and phenomenologist made this mirror apparent, so that we could begin to see how selfhood is composed in light of other persons, as intersubjective social reality.[1]

In the conversations that follow, architects describe their view through the looking glass. While each tells us something of himself, we also see the architect's portrait of our own lives. These, then, are the architect's people—actual and imagined—constructed out of personal insight, experience, self-reflection, heroes, clients, partners, colleagues, and observations of the anonymous public. It is before this living mirror that the architect practices his art.

The architects' people were quarried in May 1984, when seven New York architects were selected for interviews about the internalized advisors, dwellers, and companions that they bring to their work. These seven "star architects"[2] were chosen for their national recognition and for diversity in terms of career development, aesthetics, and stylistics. I assumed that variations apparent in their opinions and work would lead me to a varied collection of mental actors.[3] In addition, these architects are creative leaders

rather than representatives of their fellow practitioners. Another study suggests that as a group, they are more likely to bring their self-images into the
architectural endeavor than less creative architects. They are also less likely
to be guided by a sense of responsibility to their clients, the profession, or
society.[4]

Although the same questions were asked of each architect, each conversation took its own course. The subject of the interviews was generally
baffling to the architects, which is both an advantage and a limitation. At
best, complex, fresh visions emerge of architects' people as the speakers
articulate a familiar phenomenon for the first time, yet there are also places
where the architects cannot find words or images that correspond to my
queries.

The chapter is organized into two parts. In the first part, a range of
attitudes about the architect, other individuals, and society as a whole is
expressed by the architects (ordered alphabetically): Peter Eisenman, Hugh
Hardy, Steven Holl, Robert Kliment, Richard Meier, James Stewart Polshek, and Tod Williams. Their stories tell of architects' beliefs about themselves and others, and how such beliefs influence their professional activities. The second part is my analysis, organized into four sections that may
be useful to keep in mind while reading the interviews. First, the architects'
views of their own participation in architecture is described, and the relationship they envision between themselves, their buildings, and the surrounding social world. Second, characteristics are explored of individuals
other than the architects themselves and of society at large. Third, the
architects' notions are gathered about the client and a larger, more amorphous audience. Finally, an unexpected set of mental actors emerges from
the conversations: the buildings.[5]

Peter Eisenman

Eisenman Architects
Born 1932
BArch Cornell, 1955; MSArch Columbia, 1960; MA (1962) and
PhD in Theory of Design, Cambridge, 1963

"There are only four things in life that really matter: wine, food, sex, and
poetry. These are four things a computer cannot do. Architecture is a form
of poetry. These things are basic and essential. A writer like my friend
William Gass does not write for readers. He writes for himself. Architecture
is made by architects, for themselves—I do my work for me; there are no
other 'people' for the architect." In this conversation, we catch glimpses of
Peter Eisenman's thinking about himself, his work, and society. His speech
captivates, as he shouts or whispers to stage an understanding, to startle

clarity out of the ambiguity. The candor and conviction in his comments effectively provoke us to grasp his vision.

When you said you wanted to talk about "Architects' People" I thought you wanted to talk about the "me" behind the architect. For me, to *be* fully is what matters. I act through architecture. How else do I prove I'm here? Most people's lives are like being on trains going someplace, getting on faster trains, trying to get a better seat, waiting in line. I am not on any train. I live in the station between trains; the train ride is what I *do not* want. We all have tickets for the last train. The problem is, what do you do while you're waiting? I dig into the floor of the station; and this digging is life and architecture. It is the act that counts, the exploration, the mining of ideas. You only have intuition to tell you where to dig in the station. As I'm digging, I know what is dross but I'm not sure I can recognize the gold, but then I'm not looking for "it." I do not dig with a purpose; I dig for the sake of digging—it proves that I exist. What I am interested in is seeing what happens when I dig with my fingers or with a spoon, as opposed to digging with air hammers or dynamite. I like blowing things up. Nietzsche said something about destruction and creation being part of the same act. Buildings are really manifestations of this process.

There's no one behind my work but me. I am not selfish or immoral—I just want to *be* more everyday for me. To live fully, I have to uncover the Self. Like everyone, I want people to love me, to like my work, but I cannot expect it. I think people recognize that my work is architecture—they may not like it, but they still know it's architecture. And despite what people say about its abstract and difficult qualities, people understand my buildings very well. People would prefer it if my work were sculpture—my work causes a lot of anxiety as architecture because of an absence of symbolism and figuration. And then I think, "Why try to overthrow 500 years of tradition without expecting some anxiety?"

When you ask me what my approach to architecture means for the life in the building, it's one of those questions that makes me recoil. I do not know what my houses mean for living. William Gass said of House VI that he could write the most marvelous poetry, make the most marvelous love, and cook the most marvelous food in that house. Robert Gutman says that is a shambles sociologically. Well, those are two views. The human is the most adaptable of all creatures. For instance, it spends thousands and thousands of dollars to live in a very small cell on a propelled craft on the ocean, to eat in a public dining room. What does it mean for living? I'm not dealing with mindless convenience: Is the dishwasher next to the sink; is the bathroom next to the bedroom? My work is not about convenience—it is about art. I am not suggesting that people should necessarily live in art—I don't live in art—and I'm not suggesting people ought to live in my architecture. As Loos said, "Architecture is the purview of monuments and graves" (no pun intended). I could do my work without purpose—my best work is without purpose. I invent purpose afterwards. Look, the Frick Museum was not designed as a museum. Architecture added to function is adjectival, that is one of the problems. You do not read Shakespeare for the story—you already know the story. You read him for the choice of words, for the sound of the words, and for the taste of the words on your tongue. Who cares about the story? Who cares about the function? That is the reduction of architecture to mindless convenience.

I build to transcend function. The whole idea of transcending function is what makes architecture so difficult. That's what Valéry said made poetry: transcending meaning. You pick up the morning newspaper, you read it and throw it away, right? But why would somebody read it again, not for what they know but for the sound of the words? Most people buy a house because it functions to serve their purpose. How do you get them to build a house despite those needs? The insistence on the overcoming of function is the energy that begins the architectural impulse. For example, the overcoming of scale specificity in architecture, to build with scale nonspecificity, by overcoming the building's supposedly necessary relationship to human scale. My activity is not humanistic, but as Daniel Liebiskind has said, perhaps superhumanistic: an attempt that is outside the humanist discourse, outside of binary oppositions of solid and void, black and white, right and wrong, upside down, rightside up, and man and woman. But the point is not to succeed at promoting these ideas. You never succeed— you live. It is the action of architecture that matters.

Actually, I'm looking for "new readers," Foucault's new reader; I don't believe that you can read my architecture as an old reader. One would be reading the wrong things. We read architecture as speech right now. I am looking for people to read my work not as a series of images but as a reading event, as text: the idea of architecture as text; architecture that is arbitrary and without recall to type-form, or natural or divine origins; an architecture that is modificational, in the sense that its only transformation is the modification of its own structure. Its only motivation is an internal one to reach its next state of being and then it begins again—it's not going anywhere, it is just going.

I do not believe that architecture can change culture; I believe it just is. The making of it *is* it. My first three or four houses represent the life of the mind. Now I'm interested in being. I do not believe in progress. I want to do good work which follows my intentions. This takes wisdom—not knowledge, so I want to cultivate that wisdom, which is partly experiential, related to my own background, and it has to do with my capacity to invent.

To all those people worrying about making the world a better place, I say, "Why not do better architecture? Why not do better poetry?" If they would stop worrying about the rest of the world and start worrying about themselves, then the world would be *fine*. It is the people who cannot do good architecture that worry about the rest of the world. In this century, the horrors of horrors, the twin holocausts that occurred in 1945 would lead me to believe that neither science nor sociology has made the world a better place to be when man can destroy mankind at a greater rate than ever before. But the nice thing is, poets never even think about those issues, they just write their poetry. They may think about expanding consciousness, maybe, because it makes being more. I want to be more everyday. And it's enough if my work just makes me and maybe one other person be more. Somebody who can say, "I got it." I think I need one other person.

This is what I try to teach students: I try to open them up to what is being. That is, what is the difference between wisdom and knowledge. It's more like

Figure 4.1 Peter Eisenman. House X, model. (Eisenman Robertson Architects.)

group therapy. I try to help them make architecture for themselves, and to have the capacity to satisfy themselves so they will not blow their brains out when they're 35, when they find out that the two-car garage, the station wagon, the boat, the kids, the dog, the wife are nothing. We don't produce any products, I don't judge. I am, if anything, deprogramming—asking them to give up the baggage and to just *dig* with me in the station. You cannot imagine how difficult this is.

In stripping yourself of cultural baggage, what one has to preserve is the ego. That is what one is. One has to preserve one's own value, and not let it be overrun by the Self, which is certainly not what one is. The Self is the kind of demon that lives inside the unconscious. One cannot live in the unconscious, or be dominated by the Self. Even though I can free-float, and let the unconscious come up, I can also dig in the unconscious; I still have to maintain the me and know who I am. But then I look at something like this suitcase, and I don't think that an architect like me should walk around with a suitcase like this. I'll tell you a story. I was preoccupied because I was going away on business, and a friend asked me what was on my mind. I said, "Shoes."

"Shoes?"

"Shoes. I only own two pairs and they're both being repaired."

"Only two pairs? Why don't you go get one of them?"

"Well, you see, they're hand-made shoes and it takes six weeks to repair them. I send them to a special shoemaker in Virginia."

"My God, Peter. That's horrible. That's not you, that's your image of you. Tomorrow, you have to go out and just buy some shoes. Just go to the closest shoe store and say, 'Gimme those.' "

I didn't think I could do it but I did. It was unbelievably difficult. I'm trying to divest myself of images of me. This is the same process in my work.

For Eisenman, the other, in the sense of actual clients is insignificant in light of his existential project: "How do I prove I'm here?" He answers by digging, inventing, and building. However, it is against the backdrop of others, "the cultural baggage," societal expectations, that he must act in order to plumb his own depths. He does this both intellectually (overcoming purpose; building without reference to human scale) and in his everyday actions (digging, buying shoes). As he himself suggests, "I look for my Self in *your* mirror, not in my mirror. That would be narcissism." And when he talks about teaching, he implies that the appropriate human context (teacher, group therapy) allows one to recover one's identity. Thus, for Eisenman, the self is fundamentally the only member of the architect's people that can truly be known, and that is only possible with great effort.

Figure 4.2 Hugh Hardy reflects in his New York office. (Photograph by Caroline Hinkley.)

Hugh Hardy
Hardy Holzman Pfeiffer Associates
Born 1932
BArch Princeton 1954; MFA Princeton 1956

If we had to put all the architects interviewed on a continuum, Peter Eisenman might be at one extreme with Hugh Hardy at the other. It's not that they disagree—there's no opportunity for disagreement since there is so little overlap in their world views. Hugh Hardy is an energetic, practical man involved in the everyday world of experiences. "I wish we architects were more curious about other people and less wrapped up with ourselves." He is frank, quick in his responses, hates being boxed into a category, assured. Hardy prefers not to speak about people in the abstract, thinking of his "people" as those who literally participate as clients and coworkers. He makes it clear at the outset that the office's work is a collaboration and that the opinions he expresses are his own.

If there's any experience that shaped my ideas about the audience for architecture, it would probably be my work in the theater. I worked backstage, and came to realize that all those people—not just the stars—are important and contribute to making productions work. This collaboration, the humanistic exchange which makes theater, can also be present in architecture. In our practice, we don't present ourselves as experts—we don't believe in the artist-hero in that sense. The client is not a vehicle for our vision, because we believe they know as much as we do about what and why they are building. There are two things I detest in our profession: self-glorification and running down clients. I realize it's old-fashioned, but I really do think architecture is a service profession. Of course it's creative and it's an art, but it's the exchange and the collaboration that are essential to their realization.

My sense of people is not shaped by literature or social science. In school, I always found the social sciences suspect. They seemed like a device to categorize people, but they didn't go beyond that. I found myself saying, "So what?" The study of human behavior ought to be valuable to an architect, but it has not been very useful to me. The key to understanding such things is experience; I mean being curious about people, not the experience of building buildings.

Hardy parries questions about aspects of his identity that emerge in his work in part because of the importance he places on the collaboration that occurs in his office. However, he speaks freely about shared human qualities that architecture can address.

We try to make buildings that encourage you to explore. The need for discovery, curiosity, and variety is basic. Without variety you can't survive. We tried to pay attention to these things in the theater in Eugene [the Hult Center for the Performing Arts] partially by involving craftsmen and artists in the architecture [Figure 4.3]. It has given the place individuality, spontaneity, and the potential for accidental discoveries. There are cast bronze ticket stubs and gloves lying on the floor, a tile back splash for the washbasins with images that mimic your primping, and a beautifully shaped hand rail that feels good. These accidental discoveries seem to do exactly what traditional ornament does. And of course you can use the audience as a device to animate public space—there's nothing people like more than people watching. Generally, public buildings are impersonal and pompous, but I think it's all right to dance, even in public places. But you shouldn't go too far and create too much ambiguity, a disorienting situation. Take our school in Pingry, New Jersey. Its two contrasting sides ensure you can't predict one side from the other, but still you don't get disoriented. Our overall geometric organization holds the variety together with the strongest ordering devices: circulation, light, and repetition.

About clients, Hardy states, "Clients vary a great deal, but most cannot understand three-dimensional space. They can't visualize the final results, and they can't explain clearly what they want. And another problem is that people change, their values change, and ownership changes. What your initial client thought was great may be hated by the next user. That's why the architect has to rely on instinct as well as written program information."

Through the solutions Hardy feels are appropriate, we better understand his ideas about the need for variety, discovery, and spontaneity within an ordered framework. When asked about the state of the larger social context for architecture, Hardy quickly responds,

Figure 4.3 Hardy Holzman Pfeiffer Associates. Hult Center for the Performing Arts, Silva Concert Hall. (Photograph by Norman McGrath.)

The current ethos is, without question, conservative—in the sense of conserving. There's a political kind of conservation, a nostalgia for the good old days, but there's also a sense that we could all be blown up. The bomb *is* a big deal. This promotes a conserving attitude. We do a lot of work with historic buildings, like the Madison Civic Center or the Cooper-Hewitt Museum. We don't necessarily try to invent everything new *or* copy exactly what was there. This approach led to the controversy about the Langworthy house on 11th Street [in New York City]. It's in a neighborhood of row houses on the site where the underground radical group Weathermen blew themselves up. And it seemed ridiculous to rebuild that house as if nothing had happened, as if the world hadn't changed. It had. This building was criticized for having a kind of split personality, because there's this collision between one vocabulary and another. It made the preservationists angry because they didn't think the bay window belonged, and it made the modernists angry because it responded to the cornice line and fenestration patterns of the adjacent buildings, while the interior plan is set at a diagonal to the street. This controversy is related to the public/private responsibilities of architecture. The responsibility of the façade is to the public life of the street, while what you do behind the façade is your own business. Without recognition of this double burden, design is like indecent exposure. Resolving such conflicts is part of the morality of our culture.

Hardy expresses the tacit belief that buildings have a life of their own. People, to exaggerate the point, are at the behest of buildings. "In large, complex projects, no one person is in charge; you try to figure out from this consortium of people what it is that's trying to be made." At another point Hardy suggests the architects' people may stem from the images embedded in existing buildings. Calling this "the pleasures of continuity," he says,

> We live in a house in Greenwich Village that's over a hundred years old. And the way we live in it is, I'm sure, consistent with the way it was used when it was built. It's obvious my wife and I sleep in what was the master bedroom; the floor below was always the library; the kitchen was always where it is now; and the kids' bedrooms on the top floor were originally all lived in the same way. Pleasant ghosts therefore run up the stairs all the time. My daughter's violin teacher was at the house the other night and we were at the piano playing Mozart sonatas. I am sure this was not the first time for Mozart in our house. It was wonderful because we knew we were using the house in the way it was intended.

Later he describes the experience of having his own buildings demolished in his lifetime. "You almost have to consider them as impermanent objects, because in this culture, the values change so much. It's a little like parents and children. You can't possess your children, you get them launched, and then you bear with them all the good and bad of your launching, but you can't possess them."

Hugh Hardy imagines a public in whose life architecture plays an important role. Although the architect is distinct from the public, there is also a reciprocity of perspectives in this exchange. "What I like, perceive, need and so on, must be common to others. Our spontaneity, curiosity, and desire for variety can be celebrated by buildings only if others share this need. Our collective wish for a sense of security in the face of an unknown future can also be respected; our need for order over ambiguity must also be met. Architecture plays a significant role in our lives, and so as architects accept responsibility for the translation of client wishes into built reality, they must rely on their own instincts rather than fashion." All of this requires a sensitivity to human exchange. As Hardy puts it, "It is the profound interaction between self and society which both generates and illuminates architecture."

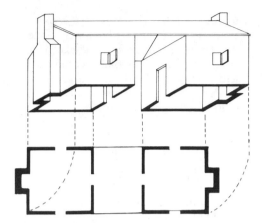

Figure 4.4 Dogtrot vernacular house form. (From Steven Holl, *Pamphlet Architecture #9,* 1983.)

Steven Holl
Steven Holl Architects
Born 1947
BArch University of Washington, 1969; graduate studies
at the Architectural Association in London

Upon arriving at the building in which Steven Holl's office is located, the telltale signs of "architect-in-residence" are entirely lacking: no white helvitica medium sign pointing the way, no luxo lamps in the windows, no pipe railings on the stair. Instead, a man with produce boxes on a dolly shows me where to whack the wall to call the elevator. The building, deserving historical monument status, appears to house potters and cabinet makers rather than architects. Inside Holl's office, two young men work amid an array of unusual models, and out the window is the building's cemetery *qua* courtyard. When Holl arrives he explains that one of the men is not really an employee but a client. It's Friday evening and no one seems in any hurry to leave.

> The private client is not so difficult, but when we talk about a public client, or mass society, we tend to become oblivious to the individuals; 10,000 individuals are not an amorphous body because individuals are idiosyncratic. I try to respond to each problem and each client specifically. This doesn't imply that experience is of no value or that my response in any particular project is entirely idiosyncratic. I think it's important to acknowledge the things that have happened in a culture prior to the project at hand. There is a cultural history that we all share, which has an underlying pattern in form. These are the typologies that I outlined in the Pamphlet Architecture series.[6] The building—the object—is a cultural fact that defies explanation. The dog-trot house, for example, can be explained in many ways [Figure 4.4]. I just read a "weather theory," and there's a hearth and spirit theory, an animals/fire/people theory, a construction theory—each one explains this basic solid–void–solid form. All these theses can be right, and belong

to the object. It's not the cause and effect that interest me, but that the center of gravity is a void. If form and function are condensed in primitive examples, we can find by metaphoric substitution not a caricature or collaged fragment but a new meaning.

Holl's people are a blend of the individual present and the cultural structures that individuals share. These structures are evident in forms, in building typologies. However, the typology is not linked to a specific function or pattern of social relations, but is open to a variety of uses and explanations. Thus, the architect's role is hermeneutic: to create an interpretation of the context (social and material) at hand, and then to find an appropriate architectural type.

For the Van Zandt house, thirteen complete schemes were developed before we showed the client even one of them. Initially, I found the requirements of having a lap pool and a jacuzzi somewhat distasteful (but you have to deal with these things), until we found this basic solid-void-solid organization, which is based on the dog trot. The program consists of a guest quarters, a lap pool, and a weekend house. The void, the pool, is the center of the composition, the focus being the landscape and enclosure bounded by the house fronts [Figure 4.5].

Nothing you end up with is arbitrary when you work like this. Take this house we designed for two artists. She paints floral, colorful landscapes; she likes cats; she needs lots of sun. He is a sculptor who works in concrete and chicken wire; he works at night and has no need for light; he hates cats. These people seemed like dialectic opposites, so we developed the scheme based on a U-shaped type, and used that plan to charge the dialectics. They have a twelve-year-old daughter who was basically outside this dialectic, so she got her own little house, a single room house with two windows and a pyramid roof.

Here Holl's role as interpreter is clear. The individuals who will inhabit the house begin the interpretive process, as Holl understands them and their relations. He finds some fundamental quality in their actions or relations and then searches for a fundamental architectural organization that corresponds.

As architect-interpreter, Holl brings the type into play when he determines it is appropriate. Thus, his own identity is unavoidably brought to the work, but Holl adds

If I base my actions, my work, on cultural notions—which no one owns—then the heart of the scheme is not ego, but emanates from the culture. I'm not talking about bits and pieces of objects. For instance, this chair: it's a version of a Shaker chair. It's not invention or caricature—it's a recalling of the original's essence. Do you know that Bartók went around Hungary collecting traditional folk songs for inspiration for his own compositions? I think of myself doing the same kind of thing. I'm trying to find a way to engage elemental aspects of the culture, its patterns. This is a kind of historical inquiry into essences, which then must be brought to the modern circumstance. Le Corbusier began to do this for

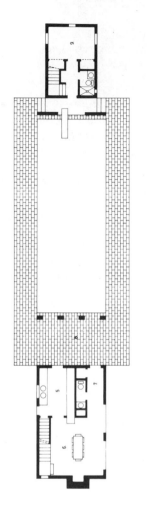

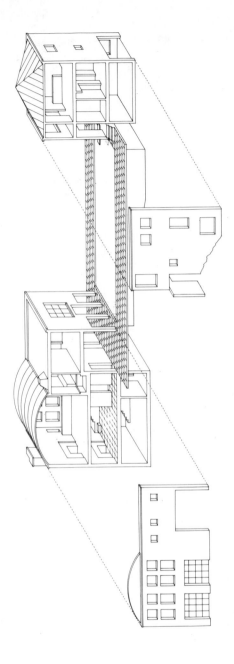

Figure 4.5 Steven Holl. The Van Zandt house interprets the vernacular dogtrot form. Ground floor plan and elevation oblique. (Office of Steven Holl Architects.)

Chandigarh in his initial sketches of the traditional Indian ways. I want to make it clear that this is not contextualism—I think that's a terrible word. The historical inquiry is not literal, since my inspiration comes from the dilemma of contemporary architecture. This is an exciting period exactly because there's no one to follow. We have to search out the means to evaluate what we're doing.

Holl has a respect for culture, for the life around him, past and present. He plans a building knowing that the inhabitants will change it to make it their own. Holl sets himself the task of capturing some essence. "Take this spec house I'm doing in Stamford, Connecticut [with the client working in the office]. The program and the site are givens. We're trying to capture the essence of dwelling—a primitive response, based on the flounder or half-house. It focuses the main rooms of the house onto a sunny rear courtyard. The site is too narrow to do this without employing some unorthodox form, and the half-house type seemed perfect. In section, it is a plan of volumes. What I would like the building to preserve, even after the dweller's possession, are mystery, layers, simplicity, and geometric rigor."

Holl's people are complex cultural and semantic entities with few behavioral or traditional psychological characteristics. This is evident in his project entitled "Bridge of Houses," which includes a House of Matter and Memory, House of the Decider, and a House for the Doubter. Holl's people have some form of an a priori essential structure that consists in part of a connectedness to cultural predecessors, and in a tendency to respond to existing (or previously existing) cultural forms. With this, Holl believes that people basically share a response to or recognition of "essences." The larger cultural context is read from the text of architectural artifacts, which offer appropriate directions for present action. The artifacts that concern Holl come from the American building tradition and the typological categories that reflect cultural patterns. The fundamental aspects of building types are distilled and reinterpreted for modern American building.

Robert Michael Kliment

R.M. Kliment and Frances Halsband Architects
Born 1933
BA Yale 1954; MArch Yale 1959

Basically what you're asking is: "Who are my companions while I work?" Clients are always my companions in a certain sense, but primarily I rely on a few extraordinary colleagues—those very few architects that I admire. At one level, there's Bob Stern and Aldo Giurgola. Stern does so many things so well—teaching, writing, building—I value the breadth and depth of his interests. And Aldo imparted certain initial ideas that both Frances and I absorbed, since Mitchell/Giurgola was for each of us our first and only job between school and the starting of our own firm. I think most architects would have to admit that they work also for the observations and approval of their peers.

At another level, there are three architects I've always seen as guides in terms of intentions and in terms of grafting to or developing from their work: Venturi, Kahn, and Aalto. I've often thought that Venturi could be the ideal progeny of a mating between Aalto and Kahn. I wonder what these three would think of what we're doing. One word of encouragement from someone like Venturi means a great deal, or also from a specialist or an informed layman.

Finally, among the most influential companions in an architect's work are his partners and associates, and in my case the influence of the companionship, both literal and metaphorical, of Frances Halsband cannot be overemphasized. Whenever I'm talking about professional or artistic issues, I'm speaking about *our* work; when I'm speaking of personal matters, I'm speaking only for myself.

Among the companions Kliment "works with," he distinguishes three different groups of people. First, with his partner Frances Halsband and with colleagues in the office Kliment shares day-to-day professional life as well as larger professional aspirations. Second, there are respected colleagues with whom Kliment consults, both mentally and in actuality; and finally, the architects Kahn, Venturi, and Aalto act as guides of the highest order.

At the same time, the best buildings come from working with stimulating and demanding clients—the ones who really make you work very hard. Just the way the client poses problems can contribute to one's creative efforts, either through conflict or through the discovery of a common purpose. Personally, I prefer the second way, because I find it hard to be creative when I'm working against a situation—against the people and their aspirations. It's very difficult to work with a client who has no interest in the aesthetic consequences of his actions. We tend not to work with them. But on the other hand, one has to educate people who may be tremendously accomplished in their own fields—which is most often the case with clients.

Kliment quickly grasps the notion of "architects' people" and describes his "companions": Some are engaged in actual dialogue with the architect, some are respected critics, and some are models. Real and imagined voices speak to Kliment, guiding him in his work. He is skeptical of broad generalizations, preferring to speak of individuals rather than beliefs about human behavior or society at large. He is a literate man who seeks the assistance and appreciation of those he most admires.

When asked about his beliefs about others, and about writers that may have helped to shape those beliefs, Kliment makes an analogy between the portrayal of life in literary structure and architecture.

I've always found the social sciences excruciatingly dull and irrelevant for our work as architects. But on the other hand, Henry James captures the subtlety and delicacy of patterns of people's behavior that one aspires to in one's work. And they are delicacies that seem logical. James' work has a seamless texture between the diagram—the intellectual and formal structure—and its continuing elaboration and refinement. We aspire to this in our office. The works of Trollope and George Eliot, especially in *Middlemarch*, may have seemed boring as a student but when one is older, one can respect how enormously difficult it is to achieve that kind of coherence. Well, I also enjoy the English public school boy adventures like John Buchan, E. Phillips Oppenheim, and Conan Doyle, and the wonderful way plot and character are woven together.

I think it's the sense of overall order in these works that I admire. I aspire to order and logic—partly because my life isn't that way. In some respects, architecture is a way to create order and logic in my own life. In my work, I *make* what I want to *be*, even though I know I'm not this. And perseverance of character, or stamina, is absolutely critical. For instance, I really admire someone like Richard Meier because his work, even though it's not like ours, is so full of determination and character.

Buildings should reflect other things that are important: They should reflect a certain grace and elegance. One likes to project into the building those things that one personally aspires to: a clear sense of order, grace, determination, maybe even some quirkiness—these are all part of character. Of course, people can do good buildings and be absolute sons of bitches, so the relationship between building and person is not immediately evident. I'm not sure all other people want these same things from buildings. There are some people who thrive on a mess, but between every architect and client, there needs to be some common sense about what constitutes order.

One makes a room that is wonderful to be in, and then every other room, every window, and so on, and in this way—every building.

In the process of trying to make beautiful rooms and buildings, Kliment feels both a responsibility and a frustration working with clients. The architect must find ways to uphold the building's character.

With a client, you have one, maybe two occasions to really misbehave and lose your temper. I suspect it's not always conscious, and one likes to rationalize in retrospect (because you hate to think that you've lost control of yourself) that it was a well-timed explosion. If you do this too often, they'll think you're crazy and you won't achieve your purpose. But when the client wants to change some aspect of the building, and you feel as if all your time, energy, effort, and money, expended on his behalf, were spent on an endeavor that's going to be compromised, then you just blow up. If you've established some trust, they'll know you don't normally behave that way, and you can achieve something by these explosions.

Kliment sees another responsibility of the architect, and another means by which architects listen to the homunculus of the larger entity: the public. "Even a private house has a public client present, to whom both clients and architects are responsible. This has to do with the pursuit of concerns that are part of the larger order of things, beyond the immediate problem. For example, the work should have some continuity with its surroundings, it should look good, and it should look like someone cared about it. Our

Figure 4.6 R. M. Kliment and Frances Halsband Architects. Entrance arcade of the Computer Science Building at Columbia University. (Photograph © 1984 Cervin Robinson.)

concerns for this public client are apparent in the Columbia building [Figure 4.6] and in the YWCA in Kingston, New York. One tries to ally oneself with the client so that he will do his duty by the public." Thus, a part of Kliment's vision of others is that they share his concerns for continuity with surroundings, aesthetics, and care.

Kliment feels that the architect has many significant responsibilities that must be fulfilled, and so the ideal architect may be the strongest human image guiding his actions. "Architecture is more than acting in this moment for one's self. Every individual, given talents and privileges, must do something to feel that he has used those gifts to some purpose. In architecture, you can contribute immeasurably to civilized existence." Kliment says that the architect's goal, in sum, is to make beautiful, carefully conceived buildings. When I ask if this means he believes that beautiful things play a central role in our lives, he replies

> But of course, don't you think so? There's both a visual beauty and a structural beauty, a certain sort of clarity and order, about the work that one does. So there

are both intellectual and visual virtues in a good building. I realize that things that look good to me may not look good to others. The issue is how connected one is to large numbers of people and the preoccupations of one's contemporaries (both lay and professional). What I was saying earlier about working for the appreciation of one's peers can lead in extreme situations to a kind of isolation from larger numbers of people, as we saw with much of modernism, at least its lesser manifestations. But finally, serious buildings, beautiful buildings, survive fashion. They can be discovered and recognized.

To Kliment, bridging the historical and formal gap between neoclassicism and modernism is an important part of the contemporary architect's charge. "It's almost as if we're trying to write a book with half a language. The best work now combines both modernism and that which preceded it, like Venturi's." For Robert Kliment, the architect brings his individual gifts and talents to the task of making buildings. The architect's personal aspirations are directly tied to the architect's aspirations for the building: Both must display character, order, determination, and grace. So that the evaluation of these aspirations and their fulfillment is not purely subjective, Kliment relies upon the informed judgments of clients and colleagues. The colleagues, heros of past and present, speak for the ideal architect who speaks to Kliment.

Richard Meier
Richard Meier and Partners, Architects
Born 1934
BArch Cornell, 1957

> "Were there any key individuals or experiences in your past that helped shape your ideas about people's relationship to place?"
> "No."
> "Or your ideas about audience?"
> "No idea."
> "What about yourself?"
> "Am I my audience? Yes and no."
> (pause) "This could be a very short interview."

In Richard Meier's office you sit on chairs he designed around a table he designed among the white models of buildings he designed. At fourteen, he decided he was going to be an architect and began preparing for his charge. Meier is a willful, thoughtful man who does not tell you what he thinks you want to hear, but what he believes. When asked for his ideas about his audience, he replies,

> Look, I have no idea who my audience is and I have no way of knowing. I'm my own audience only because I'm one of thousands of people in that audience. I'll tell you a story to show you what I mean. I was in a museum in California

looking at Etruscan heads when a woman came up to me and said, "Are you Richard Meier? I'm from Atlanta and I just want to thank you for what you've done at the High Museum." One never knows who the audience is, where it comes from, or what a thing means to them. But it's not as if the audience doesn't matter, it does. You do it for the audience—the unknown audience. You may not be able to identify them precisely, but that's who you do it for.

I do not have an image of the people that I'm building for. Imagining someone using a space is only one aspect of design decisions; more important is the relationship of any element to every other part of the building. How do you get there? What do you do when you're there? What does it mean to have it there? I realize it will mean different things to different people, so that my response and the beliefs that I hold about these things are not necessarily representative of the way others believe—I hope they are, but I cannot tell you whether they are or not. However, my role as the architect is to make this kind of judgment. My own identity, as you put it, is brought to my work, but it changes from time to time, and hopefully from project to project. The rationale for any building is always modified by aspects of personal experience in terms of the way light comes into a space, the way space is formed, molded, defined. The memory—my memory of experience—is then reinterpreted.

To understand more about Meier's notions of people, I asked about unexpected responses to his works, as a way of gaining a sense of what was intended, in the reflection. Meier searched his memory for such examples, implying that people's responses are complex and tied to circumstances beyond the building itself. Meier's unknown audience may respond unpredictably because buildings and design processes have a life of their own that is outside the architects' control.

For instance, with the High Museum of Art in Atlanta there was initially a cafe in the program. Its location and requirements influenced the design of the whole building, and then it got eliminated. These things change. The work has a life of its own that you can never predict. Again in that museum, we did an auditorium that was basically for lectures and films. After it was built, the acoustics were so wonderful that now they want to use it for musical performances. It would have been planned differently if we had known. Basically, you inject as much as you can into it and hope that is enough to stimulate it to become what you could never envision or plan.

I once did a house that I thought was the most "one-off" house, with nine or ten bedrooms, five or six baths. And I thought, if this client ever sells it, the building could be used as a small private school because it seemed too big for a house. In fact, it's the only house I've designed that changed hands. As their children grew up and went off to school, the clients decided they didn't need such a big house, and lo and behold they sold it, not to a private school but to a couple who had no children. You can never imagine what will happen over time. Everything's changing all the time, but the building still should be specific to what you perceive to be the needs and requirements of that particular situation, or its not going to function at all.

Even with public institutions where you have hundreds of people for clients, with very particular needs, the architect must try to balance these and put them into some order. But the interesting thing is, by the time the building's built, all those people are gone and the only person who's been there from start to finish is the architect. Hopefully the building is rational enough that it can adapt, within reason, to things you don't know about. The basic organization should be strong enough to hold the building—to hold the sense of spaces and hold the life within it. At the Atlanta museum, there are four quadrants around the center space, and even though the life, the organization, within these quadrants may change, the basic organization is so strong that it will always hold the building and the sense of spaces (Figure 4.7). Buildings should embody a hierarchy, a clear spatial pattern.

When asked what a client contributes to the architectural project, Meier responds,

> Every client is different, so it's hard to generalize. Sometimes it's simply a question in order to gain an understanding and that question either reinfores that which one is doing or forces one to rethink it, modify it, change it. The client's commitment to the project is essential, along with his or her faith in making a work of architecture instead of just a building. Perhaps the most important thing the client brings is specificity. You can't make architecture with nonspecificity— you can only make another building. Sometimes the problem itself is such that it limits or makes possible the quality of the work as architecture. For instance, I would never want to do a prison. I, personally, don't believe you can make architecture out of a prison since its nature is anti-life. And the site can be important—for instance, if it's underground, because light is so important. I've never seen a good subway station, have you?

Meier names a number of criteria that are essential to architecture: it must have a clear spatial organization, share some relation to the outdoors, be specific with regard to program, and hold the life within it in a positive manner. Along these lines, Meier elaborates upon those beliefs about people and their lives that influence his work. "I believe we do respond to light—to the changing colors of the day, to knowing whether it's rainy or sunny, to the changing seasons of the year. And the quality of our perceptions can be intensified and focused by architecture. We need to establish a relation between the natural and the manmade, since we all share a basic relation to nature. And we have a scalar relation to the world around us based on the nature of perception and on how big we are."

When asked if the family resemblance among his works indicates a set of beliefs about habitation, he replies,

> When someone says, "This is a Richard Meier building," I don't know what that means. People want to attach an identity to it, when in fact the building has its own identity. I may give it something, of course. The similarities among my works are because I'm interested in certain things: I'm interested in differentiating between solid and void; in focusing views from one space to another, and from

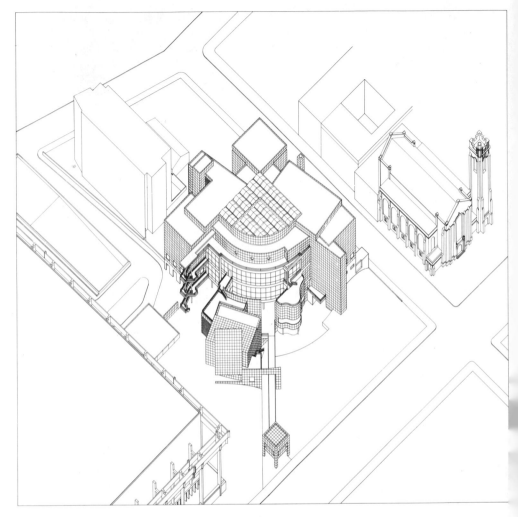

Figure 4.7 Richard Meier and Partners Architects. High Museum of Art; axonometric. (Richard Meier and Partners.)

inside to outside; in the way light comes through a building; in relationships between opaque and transparent, planar and linear, structure and skin. And I've found a vocabulary which seems to allow for certain things to happen. For example, a handrail. That doesn't change. You move along it, you hold the top of it, so in order to express that, I make linear handrails. But maybe I won't do that anymore. Some people say this is abstraction. I'm interested in abstraction because I think architecture should not be representational—of anything but architecture, that is. For instance, why should a building resemble the human figure by having a foot, body, and head? This isn't the way to organize a building.

Figure 4.8 James Stewart Polshek describes his work to the author. (Photograph by Caroline Hinkley.)

Finally, I wondered whether he felt he was making the world a better place or improving the quality of life with his architecture. "We live in a world of chaos, and we architects like to think of ourselves bringing sense, rationale, and order to this. What we try to do is uplift the spirit, and bring a sense of order into lives which without architecture might not be there. You wouldn't be an architect if you didn't have faith that you were doing something meaningful. There are lots of easier ways to spend your time than this."

James Stewart Polshek
James Stewart Polshek and Partners
Born 1930
BS Case Western, 1951; MArch Yale, 1955

Let me just start by saying that I'm not sure I have these images of people you're talking about. I think I'm satisfying an intuitive sense of the importance of the architectonic ordering systems that I see as fundamentally humanizing. In that sense, you could say I'm a physical determinist. I don't believe in user surveys

Figure 4.9 The act of a window. (Drawing by Dana Cuff.)

or psychological studies. Most people have difficulty articulating the kind of buildings they like—they take too much for granted. So I don't think you gain much by asking people what they want. I envision a solution within an instant of arriving on the site.

So begins James Polshek, architect and educator. The frankness and clarity of his reflections are consistent.

I decide if a solution is appropriate to my understanding of other people's primary perceptual responses and their memory. Memory—this is the great unexplored building block of architecture. By this I mean the complexities of the human mind, both intellectually and emotionally, which I came to appreciate when I was a premedical student, intending to become a psychiatrist. I worked at a state mental hospital before the introduction of tranquilizers and developed an enormous appreciation for the inherent order in the human mind and the role of abstraction in the absorption and reexpression of visual imagery.

That's probably why I'm skeptical about literal architectural expressions and believe that people can live and feel comfortable in "abstract" spaces. Richard Meier's buildings are no less accommodating or humane than Robert Stern's. The architect doesn't need to make the associations that people can make for themselves through their own imaginations. What architecture should respect is the importance of ambiguity, the ability to make choices in a positive way, anticipations, the organic kinship we have with a window in terms of light and views, proportional systems. There are good proportions and bad proportions. Of course there are cultural distinctions, but when you look across cultures, proportional systems always bear an approximate relation to the human figure. People remember certain experiences, and if they are frustrated or thwarted in their expectations, they become offended. In some cultures, if a window is placed so (as in Figure 4.9a), it would be interpreted as a mistake. It's almost a violent act. It invokes the memory of this kind of window (Figure 4.9b) and this kind of house, but puts them together in an antagonistic manner. To many, that would be disorienting, even anxiety producing. Or if you place a window above eye level where there is a view (Figure 4.9c)—the resulting frustration would be universal—a desire to see a view cuts across cultures. It is always important to distinguish between perceptual phenomena that are culture specific and those that are cross-cultural.

There are rules that govern the proportional systems I mentioned earlier. And I think that the successful application of those principles makes a building comfortable or uncomfortable. All architects basically agree about the Piazza San Marco in Venice, whether they're from Japan or America. The great monuments are "great" because of their common inherent properties. All the principles I'm referring to exist in a subatomic level in nature. In the end these morphologies govern so much of the logic that informs the art of building. That same logic also reinforces the principles that govern movement and perception.

When Polshek discusses the imagined inhabitants of his architecture, he makes it clear that he is not referring to specific individuals, but to his general beliefs about people.

Underlying the design of all buildings is some common sense about comfort levels. But the aspect of design that I call delight is much more complex. In the case of comfort (and here I don't mean physical comfort), all our buildings must have a logical structure and circulation system, for example. It's a three-dimensional structure that forms a matrix for circulation that actualizes the space. It's necessary for people to know where they are, where they are going, and where they have been. Orientation is a basic human need. I don't believe in making a mystery out of architecture—the mystery is in the poetics of the building—its *art*. I call the principle of circulation "the path of the feet and the eye." For example, people turn corners in particular ways, and what one sees ahead affects the way they'll turn the corner. The logic of perception and movement must be interdependent. You make things more comfortable by providing visual clues about the next move. I walk my imagination through all these situations, right down to the details of wall openings and surface characteristics.

Delight is the next level of invention. This involves taste and cultural conditioning. Delight is like clothing on the body. The bodies are all the same, more or less, but delight can differ, and you can say, "I love that color on that shape."

These notions of comfort and delight indicate that Polshek sees himself as part of the collective other, a representative to test his projects.

Architects don't see the world differently from others, it's just that they've been trained to make explicit those principles that are otherwise implicit.

In this country, where bureaucrats run things and sometimes listen to their constituents, we have an emerging pressure for democracy in design. Community design groups, review boards, and regulatory agencies are increasingly attempting to control the design of buildings. How "I feel" is becoming regulated by political forces. It's really very difficult to explain to a community planning board the basic principles of architecture, like why an entry should be placed logically in relationship to the building volume. Over time, if the building embodies all the basic qualities I've talked about, people in the community will eventually come to like it. The dangerous thing about architecture is that the most visually illogical buildings are finally tolerated. People learn to live with them. There's a natural human accommodation to the built environment once it is an accomplished fact.

Although Polshek seems to contradict himself, in fact he does not. He wants dialogue with the other, yet only if it is informed dialogue based on more than likes and dislikes. And he thinks others must respect the expertise of the architect for the dialogue to be meaningful. Levels of delight are admittedly matters of taste, but this is not equated with feelings, or offhand judgments of liking or disliking the building.

Real clients and face-to-face exchange contribute in important ways to Polshek's work.

> There's nothing more horrifying than the client who says, "Design a $2 million house for me by the time I get back from Europe." This is a private object for private consumption. For me, intellectual exchange with the client is the essence of design. And it's not enough for someone to say "it's ugly"—that's not informing. When most people come to a "name" architect for a house, it's like buying and flaunting an Yves St. Laurent bath towel or a Gucci bag. Too often wealthy men use a new house as a way to extract themselves from a marriage. In this affluent society, this is more and more common—ask the fancy decorators. I'm appalled by this. There is one type of client I do like to work with, and that's scientists. They are rational, inquisitive, and demanding. They understand the importance of first principles. Maybe it's a couple doing a house because they want a place to work. It's usually a modest little building.
>
> An office building under similar circumstances is different. It still has a public component as a piece of urban design. You can give something to the public without compromising the client's needs when you try to build buildings that do not offend the streetscape, that are good neighbors to the existing buildings. Take the example of our new convention center at Rochester, New York [Figure 4.10]. How do you break down a huge, windowless exhibition hall and make the whole thing approachable and identifiable to the public that is going to walk around it? The public circulation systems really generated the whole complex of systems that this building will contribute to, making the intersection of the river and Main Street an important place—a *memorable* place.

Generally, Polshek thinks the architect's task is most significant—most socially responsible—in the urban context with public and institutional buildings.

> I'd rather do a project like an office building than a private house. We have a project now that involves an 1,800-person suburban headquarters building for IBM—it's huge. We interviewed 80 people in depth and we learned basically three things: that they wanted to be aware of the outdoors; that they wanted more random experiences within the building; and that they wanted a more amenable and accessible cafeteria. None of this surprised us. We knew these things before and we probably would have reached the same conclusion without speaking to the employees. But we all feel better about the design this way. We approached the design as we would a small town. We created neighborhoods of 70 people,

Figure 4.10 View of the Convention Center at Rochester, New York. (Office of James Stewart Polshek and Partners.)

with specific local services. Then we agglomerated these into villages of 210 people, with more generalized services. Finally, we created a downtown with central services.

In this solution, the architect observes the human activities to be housed as a microcosm of the society at large. The labels *neighborhood, village,* and *downtown* primarily organize the population into groups who will receive services appropriate to the scale of the group. The structure of the city is adopted as a model for the distribution of services.

Polshek's humanism runs throughout his actions and opinions. The original character to inform his work is himself, typifying the complex perceiving and thinking other. As an architect, his training makes him aware of the basic ordering systems so that they can be employed explicitly, to the benefit of less aware others. Polshek does not operate on the spiritual or emotional conceptions he has of others or himself, but rather on the observable aspects of life around him, such as orientation, views, human scale. His belief in the complexities of the human mind allows him to expect more of others and to regard them with some respect. To these notions about the individual, he adds social constructs based on larger patterns of human relations that are linked to observable environments such as the neighborhood and downtown. With these images of the individual and the group, his goal is to develop an architecture that acknowledges and goes beyond his observations.

Tod Williams

Tod Williams Billie Tsien and Associates, Architects
Born 1943
BA Princeton, 1965; MFA (Architecture) Princeton, 1967.

Tod Williams' office sits at one end of Central Park in the first floor of the Gainsborough Studios, a handsome structure built at the turn of the century and one of the earliest examples of the concept of artists' housing. He takes visitors to stand in the middle of traffic so they can appreciate the beautiful double-story façade just recently restored. Inside at the back of the first floor, a room of graceful proportion that was originally the "club room" serves as the studio. Models and drawings fill the shelves, the desks, and a small skylit platform at one end. As the following discussion shows, as does all that Williams does and says, an intense commitment to architecture is obvious.

I think I had an extraordinary childhood. There's a sense of continuity, a life line and a love line, between my parents, me, and my children. As a child, I was fortunate simply to find myself around Cranbrook School, so my private, special places were also beautiful places. I still go back to them. However, like many young people, I went through some very chaotic years. There were long periods when I didn't learn very much, but somehow I was able to be touched by my environment and by some teachers in college, like Ken Frampton, Michael Graves, and Peter Eisenman. Peter does appreciate people's insides—with me he must have seen something. In any event, he helped me. Later I worked for Richard Meier, who was a hero and a mentor, but as I tried to apply that model to myself, it didn't work. That's not me. One cannot choose heroes living or dead as models for one's architecture.

Earlier in my career I believed that there was an audience out there and that the architect had some work to do. Increasingly, I realize that there is no audience and that we actually have no work to do, since architecture at its most basic level is an art and as such, at base root, it is ultimately subjective. It's not about work; it's not about problem solving as I once thought; much more it's a matter of problem stating. Understanding the relationship between the audience outside and the audience within is a basic issue. I would say that the audience is primarily within and if it's understood and listened to carefully and thoughtfully, ultimately you address a much larger and more critical audience outside. Insofar as I'm human, if I can understand my most human connectedness I would say that then I will understand myself as part of the population which not only exists at this very moment but which has always existed and will continue to exist. That's how I'm representative of the rest of the population: not in any exalted sense, but in the sense of the flesh and the spirit.

I am my audience. If you are your own audience, you have to listen very carefully to yourself, your innermost feelings, in order to know who that audience is and how to satisfy it. It's truly not something that can be described; it's something that ultimately has to be felt and reflected upon, which cannot be done quickly. It's the memory within you about how something seems. That memory can have to do with proportion, or a sensual condition, a social condition, a three-dimensional condition, a constructional condition, and I don't think it's ever only one of these things, it's a collection of different things.

I'm never without my work. It's lodged deep within me, and I have to listen to its rumblings. There's a critic within me—not in the negative sense—and as much as I can I listen to that. The critic is a kind of mirror, but ultimately creativity doesn't happen in the mirror (that would be self-indulgent or narcissistic), you actually have to do something. I believe that if you're aware of that critic in the mirror, you can do all kinds of outrageous things—truly creative and truly profound because all such things are connected with that true, real audience. I think this is related to memory—the kind beyond the mind, which is in my heart and in my gut and in each little fiber. It's completely internalized. Even in this internalized state, it must be investigated and understood, not only through feeling but also intellectually.

All work is "loaded" to a greater and lesser extent, and all work has an intrinsic depth, as do people, and so deserve an appropriate degree of attention. Issues that have to do with the spirit and with society should necessarily have more layers to them than those which have to do with commercial commodity or solely material interests. Sometimes I find myself wishing the variety of work in my office were less messy—that it all be of one type or that clients would come knowing what we might do in a certain situation. But then I would become anesthetized by a style that I might develop if I had only one type of job, or by success. I hope when I die that my work is unfinished. I'm not trying to finish anything; I'm just trying to take it in a particular direction, and what really matters is that I'm on the right track. This is extremely difficult. You could find yourself taking the easy way out, saying, "Let's do it the way we did it before," the way someone else has done it, or "Let's do it the way they want." But there's no way I will do a cartoon of my own work. You cannot stay alive that way. I want to dismantle my own principles. What must be understood is that you're building something in rather small increments and that there's no end in sight. I'd like to operate both at and beyond the levels of someone like Simon Rodia (the postman who collected the pieces for Watt's Tower as he was doing his rounds), but if I have to do it that way I can tough it out. I mean, what have I to lose? It's just my life, and it's limited.

It's hard to ask other people—people in the office and my clients—to participate in this with me, because I want everyone involved to be part of the search. Billie Tsien is my partner at the office, my wife, my right arm and my left arm, and her wisdom is the absolute best. Sometimes I think I could spare the client by just shutting my mouth and just building the damned thing. But I can't do it

without everyone's positive participation. When a client comes, I can't just pretend I'm doing something else. To do this destroys you, and then of course it
destroys the work. The trust that other people must give me is immense. Ideally,
they come because of something they believe is in the work—something smelled
right. They sniffed around and sensed a certain integrity, understanding that in
some way I was bound to the earth, on the one hand, and to the sky, on the
other. They come ready for the search. I'm not interested in only "listening to
the client": I hope I have the enough wisdom to listen to them and to me simultaneously. They bring potential to the situation and so do I, and the mix can
be very creative or chaotic.

I don't know what I'm working on right now, because, on the one hand, I
feel like I'm working on so many things and, on the other, I feel I'm not *working*
on anything at all. I just *do* it. We have a small office with a wide range of
projects. I don't really like any one of these jobs more than the other, but I'm
probably doing more work on the island job at the moment, a master plan for a
little island off the coast of Connecticut. The owner is an ideal client. He gives
me distance and respect; he challenges me sufficiently; he won't let me off the
hook, but on the other hand, he trusts whatever I decide will be right. It's hard
to live up to that, because his expectations are without limit. I care about him,
so his problems become my problems. I like the idea of patrons better than
clients. The island client is a patron. Without patrons, you're stating the problem
more for yourself; it's more self-referential.

Other people talk about the outer goals of architecture, such as public good. I
suppose that's convenient. Ultimately, there's still another one, an inner one
which I can't ignore and which I believe much more pertinent. Even though
other architects might not speak of it, it still exists. In the end, I'm the only test
of my work (I believe this to be applicable to other architects as well.) And the
client? The only way the client (and in this case, "the client" represents all outer
goals of the architectural process, including public good) won't like the process
is if I don't simultaneously listen to him or her and myself, or if I'm sloppy and
don't respect the few restrictions placed on me. As my work addresses these apparently personal issues it gets better, then architecture can become art, and then
"clients" will appreciate it for what it is.

Tod Williams speaks of himself and his work in the same breath, such
that their unity is essential to life itself. To separate one's self from one's
work is to surrender the only basis for action. In Williams' comments, work
and self are distinguished. Work is the individual's actions, but the individual consists of two parts: a present subject who acts and an entity who
guides and reflects upon the present subject. The latter, the "inner critic,"
speaks to Williams without the burdens of mundane expectations, limitations, or constraints. The present subject and the inner critic are tied to
human existence in "the flesh and the spirit," so that every action reflects
not an isolated individual, but populations of other individuals. These are
the necessary participants in the art of architecture. In terms of the interactions that occur in architecture, the ideal partner, whether client, teacher,

or coworker, is challenging, wise, respectful, and respected. Williams' own fundamental concerns then are mirrored in the other, establishing an arena for possible actions. Williams believes that some people share his fundamental concerns, and since these are essential to life itself, such concerns may be latent among others. "I absolutely believe that you constantly recover and regenerate your own spirit through your work. And when I haven't listened to myself, I have thrown it away, and away, and away, and I'm surprised I'm alive now."

Making Sense of the Architects' People

Without restating what the architects themselves make clear, the following analysis brings together individual comments on central issues of the architects' people. These issues are the role of the self, the nature of the individual and of the society, the client and the audience, and the building as actor.

The Role of the Self

From the interviews, an interesting question arises: Where do we draw the boundary between our projections of shared qualities and our projections of uniqueness? Indeed, both extremes—if I believe everyone is like me or that I am like no one else—are egocentric views of the world. While each architect draws the boundary between uniqueness and affinity slightly differently, three divisions of the world's actors can be discerned. As a whole, they see themselves among the world's actors, but with special talents and responsibilities. Naturally, each sees himself as an individual with a unique biography and set of abilities. Williams' childhood at Cranbook, Polshek's work in psychiatric wards, and Hardy's theater experience are biographical idiosyncrasies to which the architects attribute significance. Second, the architects imply that their perceptions, opinions, and actions are similar to those of other architects—contemporary peers and heroes as well as predecessors. "All architects basically agree on the Piazza San Marco in Venice, whether they're from Japan or America" (Polshek). Third, there is a group of actors who are not architects, but who are in some way engaged with buildings—as client, public, audience, or "reader." The architects attribute to this group fundamental human qualities such as a desire for variety and ambiguity, the enjoyment of people-watching, perceptual experience based on physiology and human scale, or cultural memory.

The architects speak of their work in relation to this tripartite construction of the social order: me, people like me, and the world at large, each group containing those that precede it as subsets. If architects, especially star architects, are thought to build monuments to their own egos, then these interviews provide welcome contradictory evidence. On the other hand, each emphasizes the cardinal contribution of the individual maker to the work of architecture.

We can identify several fundamentally different stances these architects take on the relationship of personal identity to the works created: self-exploration, a phenomenological perspective, and cultural interpretation. Each of these stances is a form of self-expression, if by self-expression we understand that the self is not something one discovers and then expresses, but expression is the *task* of being a united self. Architecture and the other arts are realms particularly well suited to the task.[7]

For some, architecture is inherently a self-exploration, since design offers the rare opportunity to capture a glimpse of one's self. Williams speaks explicitly on this subject when he describes his inner critic. The inner critic is the force that enables self-exploration in design: "If you're aware of that critic in the mirror, you can do all kinds of outrageous things—truly creative and truly profound." The essential self, stripped of inhibitions and reservations, permits exploration of one's full potential as an architect. As such, the creative process can be self-illuminating, but the same holds true for the architectural product. Exploring one's own identity through architecture is both to distance oneself from and simultaneously to identify with the building.[8] Meier says, "The building has its own identity. I may give it something of course. The similarities among my works are because I'm interested in certain things." Thus, if not the maker's personal characteristics, at least her or his interests will be permanently cast in architectural form. Like the results of a personality test, a building reveals a self-portrait of its maker.

A variant form of self-exploration is mentioned by Kliment, who strives to "make what I want to be"—a kind of self-construction. What he aspires toward and admires in any individual, qualities like stamina, determination, and grace, he tries to incorporate in his architecture. "In some respects, architecture is a way to create order and logic in my own life." This has a dual interpretation: the building anthropomorphically constructs the desired self; and the architect equates his own actions with attributes—I am so if I can build so. Throughout this analysis, the preeminence of action among the architects is evident. It is significant that a range of both philosophers and social psychologists have viewed actions as the basis for understanding social reality, the self, and others.[9]

Along with self-exploration, architecture is for some a way of being-in-the-world, in what might be called a phenomenological perspective.[10] "For me, to *be* fully is what matters. I act through architecture. How else do I prove I'm here?" Eisenman continues with his metaphor for life: digging in the station floor between trains, again underscoring the centrality of action to existence. Architecture, in the phenomenological perspective, is a creative, fitting domain for being. To *be* fully is the ideal, yet most difficult, way to approach architecture. Since one can never fully understand others, rummaging in the boundless storehouse of one's own nature is the only alternative. While Eisenman tends to intellectualize this process (e.g., tactical versus strategic architecture) even against his own desires, his former student, Williams, internalizes it to the point of complete identification with his work. Williams' work is in his heart, gut, and "each little fiber"; "I absolutely believe that you constantly recover and regenerate your own spirit through your work." In the phenomenological perspective, being and, in this case, the activity of architectural work are a means to uncover or regenerate some essential self that hides behind cultural baggage, inhibitions, social pressures, and other externalities.

There is a final category for the interplay of the architects' identity in the work besides self-exploration and a phenomenological perspective. Some architects express the idea that they are conduits of cultural expression, or representatives of wider social patterns. "Architects don't see the world differently from others, it's just that they've been trained to make explicit those principles that are otherwise implicit" (Polshek). In this view, architects are interpreters of society and culture. Holl takes this position: "If I base my actions, my work, on cultural notions—which no one owns—then the heart of the scheme is not ego, but emanates from the culture." From Holl's perspective, to avoid the arbitrary and the superficial, the ego must be overcome. Instead, the individual architect brings her or his abilities to the fundamental cultural project seeking cultural patterns and essences.

Looking back, cultural interpretation is the polar opposite of a phenomenological perspective, while self-exploration lies somewhere in between. Each architect constructs a perspective that integrates, rather than displays, the self with the work. A clear distinction can be made between architecture as an ex post facto homage to the architect and architecture as action that can create, uncover, and complement the self. In this vein, and parallel to the emphasis on acting and creating, the architects deny the importance of buildings as finished products. As one's identity is continually unfolding, buildings are seen as steps within the architect's life and as moments in the material history of architecture.

The Nature of the Other and the Social Order

If architects see themselves as actors in the world, the architects' people tend to be beholders. We are active beholders, however, who go about perceiving, orienting, circulating, and experiencing. We are not a contemplative lot, nor do we tend to rest between circulations. With the architect's help, we go through life seeing and orienting, yet it seems that our movement *through* space is emphasized over our actions *in* a space (e.g., resting, conversing, working). To the extent that we are spiritual beings, the architects believe our essence is bound up with beauty, art, and sometimes meaning. This essence subconsciously participates in our experience of architecture. The architects' mental navigators, by being so constructed, both reflect and create architectonic problems. Perhaps the architects' people are themselves designed to guide the architect toward the most intriguing formal problems.

Taking a closer look at the psychological and behavioral features of the architects' people, some exhibit a fundamental need to be linked to nature: to natural light, to weather, to the passing days and seasons. The link is perception, and vision is the most significant perceptual sense. The architects believe that people perceive their own scale, their corporeal structure, in relation to buildings, and while some architects try to enhance that perceived relation, others struggle to overcome it. Orientation is a basic need, founded on the common ways we move through space—or circulate. Clear circulation systems not only respond to perceptual abilities, but also to the need for order. For Polshek even comfort means "[people] know where they are, where they are going, and where they've been."

Besides the predominantly behavioral features attributed to the generalized other, less tangible qualities are mentioned with somewhat less certainty: spontaneity, curiosity, imagination, decisiveness, and desires for variety, delight, elegance, ambiguity, and beauty. These qualities contribute to the fundamental appreciation of architecture and art that is attributed to others, even if they cannot always express this appreciation. Buildings will address these qualities differently; for example, Hardy's work is almost pictorial (the tiles, bronze gloves), whereas Polshek relies on spatial ordering systems. More basic still, part of the essence of being alive is thought to be out natural draw toward beauty. The idea of an essential human nature is mentioned by several architects who seek to uncover this fundamental essence—perhaps the human analogue for a parti.

While individuals' idiosyncrasies rationalize an unknowable audience, the preceding characterizations imply that individuals are fundamentally very much alike. Some architects explain that the similarity comes from shared experience, and therefore common memories, not unlike a collective unconscious. There are experiences shared cross-culturally that come from common bonds like the scale of our bodies. For example, Polshek believes that a window above eye level, denying a view, will cause universal frustration. Other experiences are culture-specific, shared by members of some group. Thus, cultures have qualities that pertain to the making of buildings and that can be engaged by architecture. Holl thinks that a culture's existing architectural typology of forms reflects some aspects of that culture and, if reinterpreted, will contribute to it. He sees himself inquiring after historical patterns—a culture's essences.

Although the architects' people have been elaborated in terms of their characteristics as individuals, the interviews turned up little material about the relationships between individuals. Notions such as community, friendship, work relations, family, or situational encounters appear to be ill-defined among the architects. This oversight, if it is common within the profession, may have significant implications for the built environment. Besides an occasional example (Polshek's small-town concept for IBM or Hardy's own family home), the only clear references to relationships among people concern societal conditions of chaos and instability.

One of the strongest themes in the architects' vision of the social world is the need for order. Living in chaotic and troubled times, the architects' people want order and clarity in their lives and in their architecture. "We live in a world of chaos, and we architects like to think of ourselves bringing sense, rationale, and order to this," says Meier. The world is a disordered place, where frighteningly misguided actions sometimes challenge our fundamental notions of life itself (the Holocaust, the Bomb). One perceived reaction to this is a collective conservatism that is manifested as nostalgia, a desire for security, and a need for order. In this case, architecture can respond and counteract, but it cannot cure. Some believe that spatial ordering, clarity, and organization by circulation can combat the chaos. Another reaction is existential despair, as when "they find out that the two-car garage, the station wagon, the boat, the kids, the dog, the wife, are nothing." This vivid picture of contemporary existence and its meaninglessness, Eisenman suggests, can be challenged by creating architecture.

Perhaps the most interesting view of the larger social order is the notion that actors and their values are constantly changing. Not only are individuals infinitely adaptable but their expressed values, needs, and preferences are transitory. This constant state of flux makes it impossible for the architect to predict responses and thus frees the building from functional tethers. The architect must rely on instinct and intuition. Ever-changing social values require the architect to respond with a timeless, great work that goes beyond the immediate circumstance. Only Holl founds his architecture on a vision of evolution and continuity in the culture. For the other architects, the only underlying pattern of social order is the dilemma of recurrent unpredictability. The capriciousness of groups of individuals stands in contrast to the dependability of the individuals considered in isolation.

The Audience and the Client

Initially, the generalized other and the concept of an audience were equated in my own mind. After the first interview, however, I realized that *audience* carried particular implications based upon a paradoxical duality. For architects, the audience is an amorphous, unknown body of people who are touched by architecture: visitors, passersby, inhabitants. At another level, the audience is made up of specific individuals with whom some discourse occurs. It is this duality that Meier has noticed: "I have no idea who my audience is and I have no way of knowing. . . . On the other hand, it's not as if the audience doesn't matter, it does. You do it for the audience . . . the unknown audience."

An audience can be broken into constituents, among whom the most commonly mentioned is the architect: "I do my work for me," "I am my audience," "the audience is primarily within." Williams' inner critic is the means both to act and to be a spectator of that action. For some, it is through the attention to one's self that an individual ultimately connects to humanity. The architects then select who will stand beside them in their audience, including other architects such as the respected colleagues mentioned by Kliment. These are critics the architect selects for dialogue and approval, some of whom are not present or even alive. Renowned architects are invoked by contemporary architects, "I wonder what these three, [Aalto, Kahn, and Venturi] would think of what we're doing." Aalto and Kahn are "predecessors" in Schutz's terms, who can influence without being influenced.[11]

For some architects, clients are inherently part of the audience, whereas others differentiate among clients, selecting only those who are informed about and respect the architect's art. The interview structure distinguished actual clients from the idea of homonculus, primarily because most architects assume they respond not to inner constructions of social activity but to the social activity that can be immediately observed. In Mead's theory, however, both the individual's construction and the present other mingle to create the social interaction.[12]

It is clients as a group to whom architects most readily assign the category of *other*. The supposed or recommended role of clients ranges from participating in the making of buildings to participating as the viewer of an art work. Hardy says clients know as much about the future building as the architect, so that collaboration is paramount. On the other hand, Eisenman thinks people probably should not even live in his architecture, which is not about function or convenience, but art. These views correspond to the conscious removal or involvement of the self in the architectural project.

Those architects who express what I have called a phenomenological perspective are more likely to discount the active participation of the client except as a "mirror." This is seen as the existential reality, not as a denigration of the client, who by implication is only able to act similarly with the architect-as-mirror. A weaker form of the client-as-mirror is to seek out those clients whose values coincide with the architect, such as scientists (Polshek) or patrons (Williams). Another role for the client is to inspire or incite architectural quality. These clients are challenging. They make demands, they provoke; and while success is more difficult to achieve, it is also more significant. Such clients who provide specificity or provocation are conceived of as unique individuals.

Although provocative clients make inspirational contributions, others make material contributions. For Meier, architecture depends upon the specificity the client provides. His clients represent conditions similar to the site or the building materials, in that they supply the architect with boundaries and requirements. This characterization is a passive, noninteracting client who does not participate in a direct way in the building's creation. Likewise, for Holl, the client provides the material for architectural interpretation and the beginning references for a search into cultural essences.

The Building as an Actor

Architects describe their buildings as parents might describe their children: There is a degree of control by the parent-architect early in the building's life, and though the work retains these early intentions, it also changes developmentally. The better the parent-architect, the better it weathers the changes. "You inject as much as you can into it and hope that is enough to stimulate it to become what you could never envision" (Meier).

Just like the client, a building can assume the characteristics of the architect. Kliment explores this possibility when he says, "one likes to project into the building those things that one personally aspires to." From an interactionist stance, where definition of self comes largely through action, this statement reflects the architect's acknowledgment of his work's role in his life. This is similar to Eisenman's and Williams' statements that making architecture is a way to be in the world. The building then represents me both to me and to others, similar to speech. The building, like its architect, should have character (which includes order, grace, determination, and quirkiness, according to Kliment). To accomplish this in a building is to reaffirm one's own values and identity. This is not to suggest that a building must or can embody its architect's identity comprehensively. Nor does a building necessarily reflect its maker: "People can do good buildings and be absolute sons of bitches."

In this vein, the building as actor can have goodness. Good buildings are not a matter of opinion, but a matter of fact. The belief, explicitly or implicitly held by the architects, is that great architecture can be recognized cross-culturally, over the ages. This goodness is often based on the very features we attribute to the generalized other, so that a good building reflects people's virtues, or virtuous desires: discovery, variety, elegance, determination, mystery, curiosity, and beauty.

Conclusion

These architects, in sum, erode the simplistic duality between architecture as an art form and as social responsibility. The dialectic between the self and the other, the maker and the audience, the maker and the made, the architect and the client intrinsically wed one to the other. The architect's view through the looking glass to all of us and to the buildings fuses with the architect's own image to form a vision of architecture's people.

A most remarkable aspect of this study is the elaborate mental portrait of people that emerges from architects unaware that such phantom actors even exist. The tacit participation of these mental actors contributes to the shape of each building created. Varying between individuals and situations, the actors are composite bundles of architect, client, generalized other, and building. It is the architect who maintains control over the participation of each, and that responsibility is both a burden and an opportunity. The architects rely on individual experience and intuition, not social studies, when they conjure visions of situated life. The architects' emphasis on themselves as individuals may in part explain why we, the prospective inhabitants, are also seen as individuals primarily in behaviorist terms rather than as group, society, or culture.

The homunculi are difficult to keep in focus because of their kaleidoscopic transformations. Since there is a commonly held belief that people and society are unpredictable, the architect's responsibility is placed in the specific human situation at hand or in architecture's formal tradition. The supposed instability of social life leads architects to search for direction within their own souls or the building's. This is in spite of the fact that patterns of social relations like the family, the neighborhood, the work unit, or friendship are by no means elusive constructs. Although the architects deny the validity of social sciences, their own homunculi reflect the fact that architecture first brought psychologists into the fold, and only recently and in fewer numbers, sociologists and anthropologists. The connections between places and individual behavior and perception are well established, both in the academy and in the minds of designers. The weaker link is between places and groups, societies, or cultures.

Notes

1. George Herbert Mead, *Mind, Self and Society* (Chicago: University of Chicago Press, 1934. Alfred Schutz, *On Phenomenology and Social Relations*, ed. Charles W. Morris (Chicago: University of Chicago Press, 1970).
2. In his dissertation, Howard Boughey distinguishes three types of architects: hacks, stars, and ideal. The stars are artist-heroes who win awards and whose works are regularly published—a category that neatly fit these architects. Howard Boughey, "Blueprints for Behavior: The Intentions of Architects to Influence Social Action Through Design, unpublished Ph.D. dissertation, Princeton University, 1968.
3. An initial list of thirteen architects was constructed by myself and colleagues at the School of Architecture, Rice University. These architects were chosen for their national reputations, and for the differences in their aesthetics and career development. Seven of these were scheduled for interviews over a four-day period in May 1984. While the sample is not assumed to be representative of the entire architectural population, it does reflect a range of interests, stylistics, and stages in architectural practices.
4. MacKinnon's extensive study of 124 American architects examined creativity and images of self. The methods for selecting the most creative architects are similar to the methods used to select the present group. It is reasonable to assume that the New York architects in this study follow the same patterns as the most creative architects in MacKinnon's work. Donald W. MacKinnon, "Creativity and Images of the Self." in *The Study of Lives*, ed. Robert W. White (Englewood Cliffs, N.J.: Prentice-Hall, 1964), pp. 250–78.
5. The conversations were structured around four basic areas of questioning: the architect's biography and the evolution of the architects' people; the other as an individual actor (personal and psychological issues); the social world (shared attributes and societal forces); the client and the phantom actor. The tape-recorded conversations, lasting for one and a half to three hours, were then edited into abbreviated form and returned to the architects for approval.
6. Steven Holl has coordinated *Pamphlet Architecture* since 1977. He has authored four pamphlets in the series: No. 1, "Bridges" (1978); No. 5, "The Alphabetical City (1980); No. 7, "Bridge of Houses" (1981); and No. 9, "Rural and Urban House Types" (1983) (New York: Pamphlet Architecture).
7. Guy Sircello, *Mind and Art* (Princeton, N.J.: Princeton University Press, 1972). See especially chapter 10: Self Expression.
8. Lacan reasons that our subjectivity or identity is not direct but mediated by the symbolic, by which he generally means discourse. The symbolic realm inherently dissects the subject, since the symbol (in this case, a building) is always a translation of the subject (the architect).

9. Action is central to the construction of social reality for phenomenologists, symbolic interactionists, and with Lacan, even psychoanalysts.
10. Heidegger's essay "Building Dwelling Thinking" in Martin Heidegger, *Poetry, Language, Thought* (New York: Harper & Row, 1971) brought phenomenology explicitly into environmental design. Eisenman and Williams' view of architecture follows the Heideggerian model.
11. Schutz, *Collected Papers*.
12. For Mead, the construction is the "generalized other," when the particularity of the other is transcended. The transaction is between individuals, both of whom have a specific and generalized sense of the other, as well as a sense of self (the "I" and the "me"). See Mead, *Mind, Self and Society*, pp. 196–200.

SHAPES OF
SOCIAL VISION

Judith Blau, in her uniquely conceived study of architectural firms, concludes that design creativity is the singular "master value'" held by practitioners. That is, creative design is at the legitimate heart of architecture and is the professional work exclusively imbued with dignity and esteem. It would appear that the legacy of *auctoritas* and *virtu*—transformed by Wright in his conception of the creative artist, manifested in the New York architects' conversations—remains powerful among architects. Design, then, is where we must explore the role of the social in architectural imagination.

The book's first section profitably reminds us of the historic sources and contemporary expressions of values in what architects *say* about design. In the first chapter of the present section, Robert Gutman immediately reminds us that it is architecture that architects care about. We must look to buildings to understand their makers and their makers' people. Gutman accomplishes this in a lucid piece of original scholarship on the evolution of one of America's most admired buildings by one of its most admired architects: Louis Kahn's Richards Research Laboratories.

Unlike Wright, Kahn himself did not compulsively document his design thinking, but his lectures and his life as a teacher were treatises of sorts, captured by his many admiring students. Thus, it is possible for Gutman to take what Kahn stood for and compare it with what he built. He demonstrates how rich the building can be when considered as a text for understanding people—real and imagined—that live in the architect's design activity and imagination.

Robert Kerr, the nineteenth-century English architect, wrote a most extreme social document, and through a major commission, had the opportunity to realize the normative requirements of that design treatise in stone. Kent Bloomer "reads" us through the structure and the text of the "Gentleman's House," an extraordinarily articulated and rigid conception of social and physical form.

Lars Lerup finds a similar rigidity in the ossified form of the contemporary single-family house, frozen around the roles, aspirations, and limits placed on the spirit in America today. But Lerup writes as an engaged architect and liberator, choosing to display the social meaning of his own design for the problems he uncovers. Lerup finds in his examples the makings of a radical new relation between form and human activity. This chapter constructs an argument and a proposition, that architecture "must solemnly announce its own death" in order to be reincarnated without bondage to the person. If Kahn held a similar belief, we might ask whether an "unhinged subject" was also Gutman's ignored user at the Richards building.

These chapters demonstrate that the architects' people are conceived long before any individual building, so that each design solution can be viewed as an arrested moment in the evolution of an architect's social vision. Personal sociologies and psychologies are shaped, reconfigured, and learned, and from the archaeology of the designer's social vision, buildings are constructed.

5

Human Nature in Architectural Theory: The Example of Louis Kahn

ROBERT GUTMAN

Before considering the case of Louis Kahn's Richards Research Laboratories at the University of Pennsylvania Medical School, I would like to make several general comments about the use by architects of ideas about human nature. One such comment is that architecture *itself* and statements *about* architecture belong to different realms of discourse. Architectural ideas are realized in the realm of three-dimensional form. Form-making itself uses a language comprised of spaces, building elements, and materials. The statements architects make about their work are expressed in words. One cannot always be sure about the connections between statements made in the two realms. For example, are verbal statements intended to evoke a particular emotional response toward the building, or are they supposed to convey an understanding of what the building is intended to achieve? Very often a design is addressed to a relatively private audience of other architects, but to get a sponsor to finance and build it, the architect will use a completely different vocabulary to make it meaningful to the client or prospective user.

A second fact to remember is that architects borrow the concepts they use in statements about their buildings from many different theoretical repertoires. These sources must be compatible with their self-image as artists. The repertoires therefore must incorporate a view of human nature that emphasizes man's creative abilities and the unique expressive capacities of the work of art. Beyond this, however, the content of the repertoires is diverse. With respect to the relevance of architecture to what people and societies may require, the borrowed theories can emphasize, and have emphasized over the last 100 years, the human or societal needs for technological progress, for space efficiency, for visual stimulation, for a sense of community, for pride of place, for greater democracy, and for spiritual fulfillment.

Third, we social and behavioral scientists must recognize that many of our intellectual orientations are not popular among architects now. Architects are less interested in designing buildings around user requirements and programmatic concerns. These orientations are seen as manifestations of a positivist and empiricist bias. The theories of the social sciences that have intrigued architects over the last decade are traditions that emphasize the universality of social structural and mental forms (e.g., Gestalt psychology and structuralist thought) that investigate the role of symbolism in culture and society (e.g., symbolic anthropology and religious sociology), or that examine the impact of social change on culture (e.g., Marxist humanism and critical sociology).

Fourth, in reflecting on the ideas architects have of people it is also important to recognize that people and their satisfactions are not the primary concern of most architects. The principal interest is architecture, and architecture, at least in its manifestation as an art, is believed by most advanced architects to exist in a realm by itself. Architects know that the practice of their art has a closer functional relation to people than painting and the other visual arts. Without a sponsor a building does not get built. Also the very nature of architecture as an art form requires that it provide usable spaces for individual and group activities. Indeed, there is now a well-established tradition, which began with Pugin, that argues that the test of a building's aesthetic qualities is its effect on social relations. Nevertheless, the main thrust of architectural endeavor, the subject matter of architectural theory, has been architectural form itself. I do no wish to overstate this emphasis: The evaluation of form by the designer and the justification of it to other architects and to the community at large have involved discussion of user requirements, but this attention to human or social implications is usually an ancillary interest of the designer.

The balance between the attention to the purely architectural and a concern for usability seems to change every few decades. Le Corbusier and Gropius used the language of social and political thought to formulate their program and incorporated it into their architectural theory, but ultimately their forms were based on architectural typologies. In recent years, under the banner of postmodernism, architects have exhibited less of a sense of obligation to claim that the buildings they design have a moral or social content and are more frank about their inclination to tailor social and political ideas to their architectural ambitions.[1]

Fifth, it is useful to remember that architects are not, generally speaking, systematic social or psychological theorists. This follows in part from what I have just said about the primacy of the design realm. Architects are brutally eclectic in their choice of theories and do not appear bothered by contradictions in their belief systems. In this sense, they are like intellectuals in other fields. Most of us are systematic and precise in our thinking only in those areas to which we have a professional commitment. Eclecticism in social theory runs deeper among architects, however, than in many other professional groups, again because their basic commitment is to the visual realm, and statements composed in words are therefore less crucial for the success of their imaginative enterprise.[2] Another manifestation of eclecticism is the tendency of architects to leaf through books on social science and philosophy, looking for phrases that express their personal views and lend an imprimatur to their design work. This habit probably is the extension into the verbal mode of the ability to make rapid evaluations of the quality of design work.

The final point I wish to make about the role of social theory in architectural design is that many architects believe that production requires concentration on issues that are important to an audience of fellow architects, and to this audience only. The articulation of this viewpoint dates to the 1960s. It is well represented by a passage in Robert Venturi's manifesto *Complexity and Contradiction in Architecture,* in which he called on architects to narrow their concerns and concentrate on their own job of making architecture.[3] Carried to its extreme, as it is in the hands of the postfunctionalists, the search for autonomy turns into a hermetic view of architecture. Architecture can only be about itself because, in this faction's view, the "present has no future, and all that is left is to make empty words."[4] Their apocalyptic interpretation of the fate of architecture is, to say the least, unusual, but it does reflect a sentiment that makes comprehensible the lack of interest of many designers in "people" and their needs.

It would be grossly unfair to Louis Kahn to suggest any affiliation with the postfunctionalists. His socialist and New Deal leanings and experiences during the 1930s and 1940s suggest that although Kahn is often regarded as a precursor of postmodernism, he was also close to the humanitarian concerns of the modern movement.[5] Nevertheless, it is true that the depth of Kahn's involvement with issues that were of major concern to the architectural discipline after World War II led him to concentrate on certain formal moves. These were pursued with such integrity and originality in the Richards Research Laboratories that the building became part of the canon of postwar architecture before it was occupied. However, his drive to address these issues was so intense that Kahn was unable, at least at this point in his career in the late 1950s, to respond to the legitimate and genuine fears of the scientist–users that the building would impede and thwart, rather than enhance, the progress of their research endeavors. One of his admirers, the English architect Peter Smithson, has summarized the ambiguity of Richards as follows:

> Kahn's work might be regarded as the personal attempt by an architect who believed in formal order in the classical sense to come to grips with the problem of accommodating the criticality and universality necessary to contemporary architecture. This attempt or search was the result, one feels, of his realization that neither the universal space nor the space precisely fitted to function was entirely satisfactory, and that design had come to a dead end. . . .
>   The Richards Medical Building, despite his efforts, was nevertheless far from successful as a place to do research. . . . It was an inappropriate container for the various conditions and demands, both predictable and unpredictable.[6]

The Richards Research Laboratories is a building of approximately 90,000 square feet, organized into four connected towers of eight stories each (see Figure 5.1). Three of the towers contain space for laboratory work with the fourth tower housing animals used in biomedical research. The building is fully serviced with the utilities needed for modern biomedical research. The original occupants of the building were the departments of medical genetics, physiology, microbiology, and surgical research of the University of Pennsylvania Medical School. Kahn received the design commission in the spring of 1957, ground was broken in the late fall of 1958, and the building began to be occupied in the winter of 1960–61.[7]

Richards had six features that lent it distinction as a work of architecture, and some of them were directly responsible for the building's "failure." The parti was the most revolutionary of Kahn's proposals. With the exception of the Johnson Wax tower, designed by Wright and completed in 1950, scientific research facilities until Richards had been big rectangular boxes.[8] In these standard buildings, the laboratory equipment and spaces in which scientists and their assistants carried out inquiries and experiments were located in the central area of the building, along with the ducts and shafts for support services: elevators, stairways, piping for the delivery of water, heating and ventilating ducts, and the returns and flues. The offices in which the scientists wrote their papers, conferred with colleagues and students, and did their administrative work were generally located around the perimeter of the building. The corridors were positioned between the offices and the labs. Several variations on this basic plan were developed, but the organizing principle of labs and a service core on the inside, then corridors, and offices on the outside was more or less standard.

Kahn, as is well known from the Richards, from his writing, and from his later project, the Salk Institute, was unsympathetic to the box plan. In reaction to the separation between working space and corridor, Kahn de-

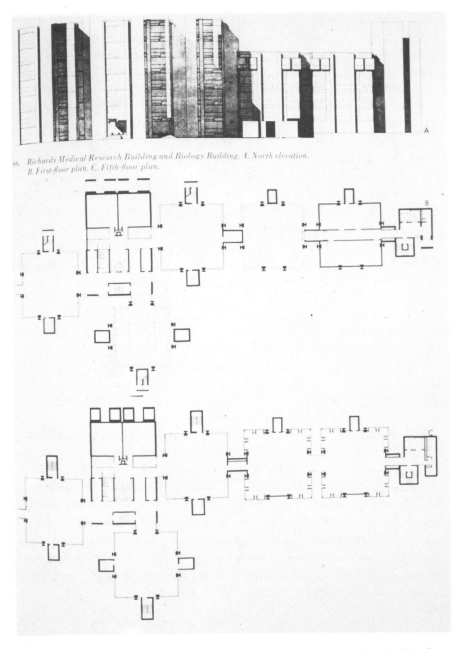

66. *Richards Medical Research Building and Biology Building. A. North elevation.*
B. First-floor plan. C. Fifth-floor plan.

Figure 5.1 The official drawing of the Richards building, showing the north elevation, the first-floor plan, and the plan of the fifth floor. The original of this drawing was exhibited at the Museum of Modern Art in the spring of 1961. The three towers to the right of the cluster of four make up what is known as the biology building. (Kahn Archives, University of Pennsylvania.)

signed a building that had no corridors. The whole of each floor of the "studio" towers was work space, and it was up to each group of users to designate areas for circulation. Similarly, within the 45-foot square of each studio, laboratory and office areas were not differentiated: the entire studio was conceived as a single, unitary space. Support services were removed from their hidden locations in the building core, moved to the outside, and placed in the famous "service" structures running vertically alongside the studio towers.[9] Sheathed in brick, some were stairways and others contained pipes and flues. The services were also made visible inside the building, as they passed under the ceiling through an open plenum.

Although there were rumblings about the tower form from the time the project was first unveiled to the medical school planning committee, the physicians and research scientists concentrated less on the shape of the space than on the amount each department would be assigned (see Figure 5.2). Their complaints about the horizontal arrangement of the lab space on each floor came later, after they moved in, because of the difficulty of expanding a laboratory experiment once it had used all the area assigned to it within a studio. The absence of designated corridor areas could, in theory, be helpful in this respect, but it often conflicted with the need of the scientists to move into and out of the areas between towers. Trying to find circulation space was a bewildering experience, even for people who worked there a long time. Perhaps the best evidence for the essential impracticality of the arrangement is that Kahn abandoned this feature of Richards when it came time to design the Salk Institute.

A second distinctive feature of Richards was the laboratory spaces themselves. In Kahn's terminology, they constituted the "served" or working spaces of the building. Perhaps the most important concept that defined the studios is that they were intended to be kept unpartitioned. The process through which the open studio concept developed in combination with the idea of the tower form is unclear, but as an idea it obviously touched Kahn's very deepest beliefs about architecture. For example, the unity of the studio space was connected to Kahn's conviction that the "room" was a basic element of architectural composition. It was a theme in Kahn's rejection of Miesian space concepts, about whose deficiencies he could speak eloquently:

> You should never make a space between columns with partition walls. It is like sleeping with your head in one room and feet in another. . . . Space is not a space unless you can see the evidence of how it was made. . . . What I would call an area, Mies would call a space, because he thought nothing of dividing a space. That's where I say no. . . . In the Miesian space he allows division, but for me there's not entity when it is divided.[10]

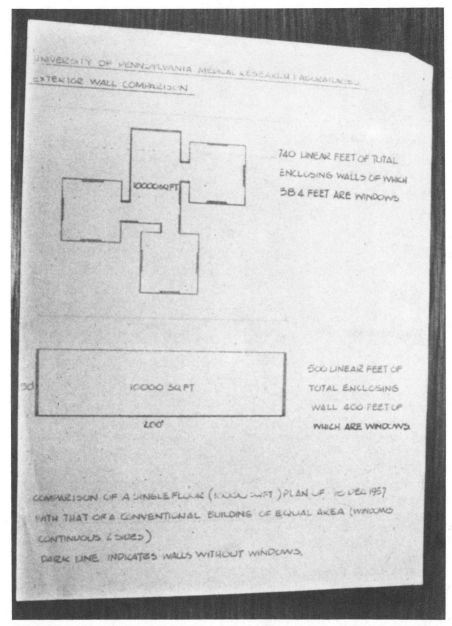

UNIVERSITY OF PENNSYLVANIA MEDICAL RESEARCH LABORATORIES

EXTERIOR WALL COMPARISON

10000 SQ FT

740 LINEAR FEET OF TOTAL
ENCLOSING WALLS OF WHICH
384 FEET ARE WINDOWS

10000 SQ FT

200'

500 LINEAR FEET OF
TOTAL ENCLOSING
WALL 400 FEET OF
WHICH ARE WINDOWS

COMPARISON OF A SINGLE FLOOR (10000 SQ FT) PLAN OF 10 DEC 1957
WITH THAT OF A CONVENTIONAL BUILDING OF EQUAL AREA (WINDOWS
CONTINUOUS 2 SIDES)

DARK LINE INDICATES WALLS WITHOUT WINDOWS.

Figure 5.2 A mimeographed sketch prepared by the Kahn office in December, 1957, comparing the plan of Richards with the standard box design of the typical laboratory building. The sketch was developed to answer critics of the Kahn design by the University Buildings and Ground Committee. Some committee members wondered whether the Richards parti had a larger perimeter and therefore would be more expensive to construct. Although Kahn's reply was not really satisfactory, the committee approved his proposal. (Kahn Archives, University of Pennsylvania.)

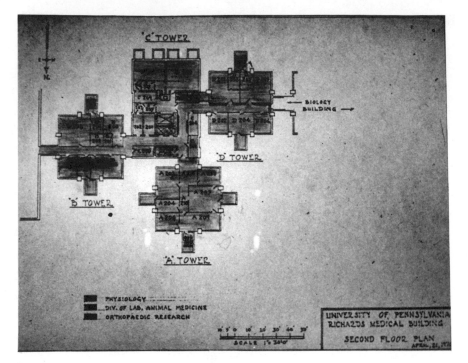

Figure 5.3 A drawing prepared by the university planning office showing the partition system and the pattern of use on the second floor of Richards in 1971. The second floor was occupied mainly by the physiology department. (Kahn Archives, University of Pennsylvania.)

His intellectual ties to Beaux Arts functional theory led Kahn to offer *quasi-pragmatic* arguments as well. For example, Kahn believed that separate studios would help scientists—who, according to his observations, generally worked as a team—to experience a sense of their group identity. He implied that this would not happen in the indeterminate spaces of conventional laboratory building. "A man may work in his own bailiwick," is how he described life in a Richards studio, compared with the feeling of a lab or office in an ordinary building. In the standard box plan, he wrote, "the only distinction between one man's space of work from the other is the difference in the numbers on the doors."[11]

Kahn's hope that his building would encourage the use of the lab space in the way he envisaged was not confirmed in practice.[12] The open studio encountered more opposition from the future users than any other single feature of the design. It was shot down almost from the day Kahn proposed it. When the first drawings were submitted by his office to the medical school administration in the late autumn of 1957, they were distributed by the dean to department chairmen with instructions to mark the prints to indicate where the scientists wanted their office partitions located. "Walls can be placed along any of the gridlines or within the corridor-type [*sic!*] blank spaces which extend the full width and breadth of the diagram. . . . Please feel free to work directly with Mr. Kahn or with Mr. Vreeland [Tim Vreeland, an associate from Kahn's office] as necessary."[13]

The physician–researchers objected to the open studio because they preferred privacy (see Figure 5.3). Instead of exhibiting the generous spirit and altruism with which Kahn aspired to endow it, the scientific enterprise is highly competitive. The average scientist does not like to reveal discoveries before his or her claims to authorship have been clearly acknowledged by colleagues.

It is interesting in this regard that the few lab spaces of Richards that were kept relatively open were all in the studios occupied by the biophysics department. In biophysics most of the scientists were responsible to a director who selected the problems they should work on, in contrast to the practice in other departments in which senior staff and faculty determined their research topics and work habits themselves. One of the features of the open studio that appealed to the director of biophysics was that Kahn's design made it easy to oversee the activities of department staff (see Figure 5.4). The director located the space where he carried on his own research so that it, too, was always visible and accessible to junior colleagues. Indeed, this space became a much-used circulation path by staff members, a development the director encouraged. However, it is ironic that Kahn, an advocate of freedom and autonomy for artists and scientists, should have used a laboratory form that facilitated a bureaucratic rather than a colleagual type of research organization.

The open plenum was another feature of Richards that met heavy resistance before the building was completed. Drop ceilings were being added in in the spring of 1960, eight or nine months before the first group of scientists moved in. Kahn indicated his aversion to hung ceilings on theoretical grounds when he designed the tetrahedronal ceiling of the Yale Art Gallery in 1953, but in Richards he was forced to modify this principle. The biologist who was offered the position of chairman of the microbiology department, the scheduled tenants for two floors of Richards, insisted upon the change as a condition for accepting the job.

The combination of partitions and drop ceilings was disastrous for ventilating, heating, and air conditioning the building. The major reason the problems arose was that, as Kahn's office explained to university officials in December of 1957, "Each floor of each tower is considered as one temper-

Figure 5.4 A studio area in the biophysics department on the fifth floor of Richards in 1972. Most of the plenum in this studio remained open just as Kahn designed it, although in other departments drop ceilings had been added by this date. (Robert Gutman.)

ature zone."[14] Several change order adjustments were approved and carried out under the direction of Kahn's office when the hung ceilings were added in microbiology during the spring and early summer of 1960.[15] Departments assigned to other floors also added ceilings after they occupied the building in 1961, with money taken from endowment funds under their control. However, the technicians who were employed by these departments did not alter the ductwork. As a result, in many areas of Richards, although there were ducts to deliver the warmed, and later on, the cooled air to different locations on each floor, the only air return was a central collector located near the elevators. It is doubtful this arrangement would have provided an adequate number of air changes, even if the partitions and hung ceilings had not been installed.

The chairman of a department that tapped its private endowment to install drop ceilings and whose personal lab and office were filled with fans and space heaters in an effort to improve indoor air quality, summarized his feelings about Richards in a letter to me. I had written him requesting an interview.

Dear Dr. Gutman:

I suspect that nowhere in the history of 20th century architecture could one find a better example of an edifice which, over the same interval of time, has:

> i) Enhanced the stature of its creator in the eyes of his profession (students of architecture swarm around this building like Beatle Fans at a rock festival) and
> ii) Seriously impeded the progress of medical science, because of its gross inadequacies from the viewpoint of those who have to use it.

Of course, as a tenant of five and a half years' standing in the Richards building, I'll be happy to meet with you. We can even discuss the problem of disassembly and crating for transmission to the Smithsonian Institute.[16]

The letter's author, a fellow of the Royal Society, left Penn for a job at another university. He told me that one of the reasons for his departure was his frustration in dealing with the problems of Richards. Even if this was only an alibi, in the 1980s the university began to renovate the HVAC system, finally adapting the ductwork to the partition and ceiling arrangements that exist in the building.

Several other features of Richards were as important as the parti and the concept of the open studio when considered from the point of view of the building's innovative role in the history of postwar American architecture. None of the others, however, met the same degree of resistance from the users, at least during the design phase of the project. For example, the constructional system was considered innovative in 1958 and was also an important manifestation of Kahn's theories of architecture. Despite the fact that the use of precast structural members increased the cost of the building over poured-in-place concrete, and thus may have been a source of the many revisions required in the plans before the project got going, the reviewers appear not to have singled this feature out for blame.

The provisions in the design that allowed natural light to enter the building were also very important to Kahn. "No space, architecturally, is a space unless it has natural light," he said. "A room in architecture, a space in architecture, needs that life-giving light—light from which we were made."[17] Applied to the plan for Richards, this principle resulted in fenestration on all sides of the towers, except where the towers joined the central core of the building. The windows were big in relation to the area they were supposed to illuminate, stretching from the spandrel at ceiling level to the brick wall at workbench level. The windows also became progressively bigger as the studio reached outward, and were tallest where they joined the window that formed another side of the tower. This last feature of their design was questioned during the reviews of the building by some university committees because of the window's appearance, not on the pragmatic grounds that the design let in more light where it was least needed.

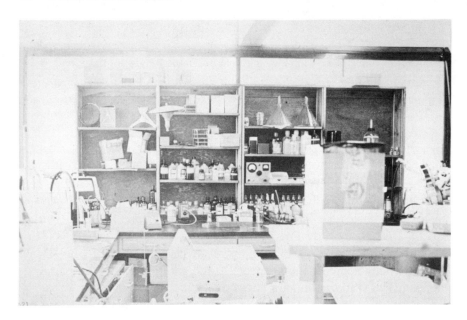

Figure 5.5 Glare was and remains a common problem in Richards. This illustration shows one of the many ways in which users controlled the offending light level, at the same time adding some much-needed shelf space. (Robert Gutman.)

Considering that distress over the intensity of the light in Richards is probably the "failure" best known to scholars and students of the building and the one most often photographed (obviously because it can be seen from outside), it is interesting that there is so little evidence from the history of the development of Richards to indicate that this was a major concern of the clients and prospective users (see Figure 5.5).[18] One reason is that the users were enchanted by the prospect of laboratories illuminated by natural light. They were also attracted by the thought of the views of the Penn campus they would see from the upper floors. The views were marvelous, because of the many vaulted and domed tops on buildings at Penn, and thus constitute one of the great amenities provided by Richards. Another reason for the initial absence of distress over the large windows is that Kahn's office put considerable effort into trying to find a screening material that would admit the full volume of natural light when it was wanted, would not obscure the panoramic views, and still could be used to modulate the light when research required it or when the thermal loads were heavy during a sunny day. However, the available window technology did not do the job adequately, including a material, Koolshade, which Kahn's office believed to be capable of solving the problem. The client asked for its removal six months after the building was opened for occupancy, apparently because it interfered with the scientists' views of the campus but did not accomplish its purpose.[19] It was at this point that the old newspapers, aluminum foil, bamboo shades, and vertical blinds began to appear all over the building, on the northern as well as on the other exposures.

Medical school officials adhered pretty closely to the original budget projection, if anything paring it somewhat as the process went forward. There were two reasons for their position. First, they had a promise of $1.5 million toward the total project cost (including fixed and movable equipment) from the National Institutes of Health, which had been expected to cover one half the expenditure on the building. The Penn administration looked forward to getting more such grants in the future and was anxious to maintain the university's credibility. Second, the original $3 million was a total project cost, including equipment. As the day when Richards would be finished came closer, the medical school officials became aware of the price of the equipment that had to be installed, so they tried to cut back on the cost of the building itself.[20]

The record of the negotiation between Kahn and the client gives the impression that Kahn hoped that the budget figure would gradually expand as the client became more entranced by the building's design. This is not to say that Kahn was irresponsible or unresponsive. He was quick to try to modify the designs when the estimates proved higher than expected, always, of course, pressing to keep his basic design intact. His difficulty in accurately forecasting the cost of the building reflected his lack of experience and that of his office colleagues in dealing with a project of the complexity and scale of Richards. It also resulted from the uniqueness of the constructional system and in part was the consequence of the pace of technological development in postwar medical science. Penn officials and the medical school faculty were not that much more adept than Kahn when it came to thinking through the requirements in advance.

From the perspective of someone situated outside the world of architects, it may be surprising that the innumerable problems that the client and users have had with the Richards building have not shaken the architectural community's attachment to the building or diminished in any way their reverence for Louis Kahn. I think this assessment can best be understood by what I said earlier about the separate realm in which architecture and the architect dwell, not only in terms of discourse but also in terms of standards of building evaluation. Kahn suffered personally in the sense that as a result of Penn's problems with Richards, he was taken off the list of architects considered for new buildings and never was given another design job on campus. When renovations began at Richards, Kahn's firm bore the ultimate indignity for an architect: other Philadelphia firms were brought in to make the alterations to his building.[21] Kahn's success in producing a building that is a landmark of postwar American architecture and the problems that led to his not getting another commission from his beloved alma mater are related events that stem in part from his ideas about people and the way those ideas were used in the design of Richards.

The use of the term *people* to denote the individuals or groups who inhabit or use a building comes out of a specific theoretical tradition insofar as it relates to architecture. It is the tradition associated with the behavioral science approach to architecture, in which it is assumed that the person occupying a building knows what he wants and that the test of a building's adequacy is whether the building design and the building itself conform to these expressed requirements. As I indicated in the first part of the paper, this approach to the human aspects of architecture is not very popular among architects now. Kahn was one of the prominent American architects involved in the movements that sought for other ways of thinking about individual and group needs in relation to architecture. It was Kahn, for example, who in reaction to the rise to prominence of the positivist approach to architecture in the 1960s emphasized the distinction between needs and desires. Architecture, he said, should aim to respond to desire.[22] Kahn told his young colleague and disciple, William Huff, that the reason he regarded Jonas Salk as his best client was that, "He did not know what he wanted, but he knew what he aspired to."[23] The Richards building was conceived in an attempt to make these aspirations "visible," but unfortunately for Kahn, the client was less sympathetic than Salk.[24]

Kahn's ideas about human nature and its implications for architecture evolved over a number of years. He stated them publicly for the first time in lectures given during the 1950s, and they were not fully articulated until the few years before his death in 1974. The earliest versions of his personal philosophy are contained in remarks he made at Princeton in 1953 at a conference on architectural education, when he was 52 years old; and in the letters he wrote to his friend and collaborator Anne Tyng expanding on these remarks.[25] In this statement Kahn emerges as a functionalist in the tradition enunciated by Sullivan. As Norris Kelly Smith as pointed out in discussing the theories of Louis Sullivan, the purport of the phrase "form follows function"

> is not that the form of a building should logically be derived from, and only from, utilitarian and structural considerations, but rather that it should exuberantly proclaim, should radiantly show forth, the goodness of the human experiences which the use of the building will give rise to.[26]

The tower form and the glass-enclosed studios that gleam in the reflected sunlight and when illuminated from inside at night proclaim an idea, make the Richards a perfect representation of Sullivan's view of functionalism. They proclaim the nobility of the community of science. In Kahn's hierarchy of requirements the celebration of this idea took precedence over such considerations as environmental standards for bacterial and viral research or the need to conduct experiments free of excessive heat and glare.

In searching for a theory of human nature that would support his version of functionalism, Kahn, like Sullivan before him, had recourse to Neoplatonic idealism. Neoplatonists believe that the only reality is mental states, such as ideas and concepts. They do not deny that there is a material level of existence that is presented to the senses, but they claim that knowledge thus derived is often deceptive. Truth lies above and beyond this material existence, in some transcendent realm that we can perceive only indirectly and with difficulty.

Kahn's commitment to philosophical idealism is probably most clearly expressed in his famous phrase "what the building wants to be." Many people have interpreted this as a version of anthropomorphism, as if Kahn were attributing to the brick, concrete, and mortar themselves an animate power leading toward a certain form. A more accurate reading of what Kahn is saying is "what the idea is which wants to be manifested in the building." This helps us to understand Kahn's definition of design and its relation to the idea of form.

> Design is a circumstantial act, how much money there is available, the site, the client, the extent of knowledge. Form has nothing to do with circumstantial conditions.[27]

Form may be regarded as the physical manifestations of the idea that is essential to the building, an idea that grows out of the architect's understanding of the nature of the institution that is the true client for the building.[28] Design, in contrast to form, is the specific architectural idea or "construct" in terms of which the order or form is manifested and is, as Kahn's statement indicates, a response to specific program constraints. In the case of Richards, the form of the building would include the studio plan and the relation between servant and served spaces, but the design encompasses such specific details as the number of stories, the use of brick rather than concrete, and the allocation of one tower to animal labs and three to research labs.

Kahn's interest in Neoplatonic idealism exhibited a specifically architectural twist. Like Sullivan, he believed that architecture had the capacity to manifest transcendent ideas and, in doing so, could bring the way of life implied by the ideas into being. In a talk he gave at Rice in the late 1960s, he said,

> I know of no greater service an architect can make as a professional man than to sense that every building must serve an institution of men, whether the institution of government, of home, of learning or of health and recreation.
>
> One of the great lacks of architecture today is that these institutions are not being defined; that they are being taken as given by the programmer and made into a building. [29]

A big difficulty in the path of achieving this goal was that clients had lost touch with their aspirations, because they were corrupted by the prevailing social institutions that were "rotten to the core." [30] It was his job as an architect to liberate them from their own materiality. Although inevitably this view cast Kahn in the role that today we might denounce as elitist, the label is irrelevant because it ignores Kahn's utopian intention. He was struggling to realize an idea whose end result would be an improved quality of life for the scientists, and perhaps increase their productivity. Kahn would have been overjoyed had the scientists seen the point themselves and adopted a strategy for creating a unified community of research workers. But the users of Richards did not think along these lines, either because the issue did not interest most of them or because they lacked the capability within their own organization. He reported proudly to Kommendant that the Penn scientists asked him many more questions than he asked them and that for the most important question he put to them—"What *is* a medical laboratory?"—they had no reply. [31] In the absence of a clear answer, Kahn seized the opportunity to provide it according to his personal interpretation of their needs.

To argue that Kahn's definition of architecture was rooted in Neoplatonic idealism leaves open the question of why the use of that orientation should have given rise to a building form centered at least partly around the concept of fostering a *community* of scientific workers. I believe the answer lies in an attitude Kahn shared with other American architects in the romantic tradition, particularly Sullivan and Wright, who were rebelling against the increasing impersonality and anonymity of industrial society. For example, in his plans both for the Larkin Building of 1905 and in Johnson Wax during the 1930s, Wright was promoting communities of workers who would be inspired to achieve new levels of production combined with a sense of social responsibility and public service. [32] Kahn's laboratory studios, which he sometimes referred to as sacred places, would, he hoped, contribute to the achievement of similar goals.

Given a Neoplatonic architectural theory, there is the question of how the architect deals with the specific user requirements that may come into conflict with a building that attempts to realize the idealized version of a program. Kahn was convinced that the way to proceed was to redefine the program, making it compatible with the idealized version, thus also providing a means for transforming it into architecture. The user's program is too scattered to be made into architecture; the version that is expressed in terms of an idea at least might be. Kahn expressed this conviction on a number of occasions, of which the following examples are typical.

> I believe it is the duty of the architect to take every institution in the city and think of it as his work, that his work is to redefine the progress brought by these institutions; not to accept programs but think of it in terms of space.[33]

> JC: Let's say you are given a commission, a specific project. Do you begin as if you had no client—with dream drawings fresh out of desire?

> LK: It must begin there.

> JC: Must every project begin with that kind of . . .

> LK: Absolutely. It must begin without a client, because the client must not order.[34]

Even if the architect succeeds in his search to redefine the program to make it compatible with the form around which he hopes to develop the design, he still must persuade the sponsor to approve it. Kahn was very alert to this problem. One should not assume just because he had a ruminative personal style that made him appear remote from worldly concerns that Kahn was naive about he difficulties of dealing with clients. He was remarkably skillful in setting forth his ideas in a way that was attractive to some users, and this undermined the effort of other users to get him to reverse or modify his proposals. I noticed in my own conversations with him, and in my observations of his encounters with others, that he could discern very quickly who would and who would not be receptive to his speculations about the meaning of life, society, and the experience of architecture. When it was useful to the achievement of his architectural aims, Kahn could speak the language of functional requirements and space planning. But he much preferred a client or situation that allowed discussion of questions that transcended the practical realm and that constituted, for him, the essence of architecture. In this respect too he was akin to American architects such as Wright, who typically accompanied their design proposals with fulsome explanations of the building's implications for transforming the life-style of the inhabitants and users. It is said that Salk sensed Kahn's interest in making buildings that would transform the way in which science was conducted in the typical laboratory and selected him for the project in La Jolla in part because of it.[35] Early in the design process for

Richards, Kahn identified scientists such as the head of the biophysics department who were on his wavelength. He also maintained regular contact with his major patron and supporter in the university, the new dean of the Graduate School of Fine Arts.

The different measures Kahn adopted to control criticism of the building were not sufficient to transform the negative evaluation by the medical school administration and the users. Indeed, the attacks became more widespread following the full occupation of the building during January and February of 1961. His defenses were initially pragmatic. For example, he and his office colleagues made many visits to Richards, attempting to adjust the heating system and the utilities needed for research, making further modifications in the partition system and the furniture, and testing new screens to deal with the problem of glare and the penetration of cold air through the windows. These "shakedown" problems are standard in new buildings. In the case of Richards, however, the operational difficulties multiplied, and the criticism did not abate. In fact, they escalated, as structural failures developed in the corners of the studio floors, because of a design failure in the truss system. Kahn was still involved in the design of the second stage of the Richards, known as the Biology Building (see Figure 5.6). The university administration decided to relieve him of further responsibility for this job and called in an engineering firm to complete it. [36]

In view of his lack of success in mollifying the Penn administration and scientists, it is perhaps not surprising that Kahn adopted a psychological defense to substitute for pragmatic measures. Although he knew, for example, *three years* before the building was completed that the studios would be partitioned and drop ceilings added on several floors, he described the building in terms of the studio concept and the open plenum whenever he gave lectures about Richards, even many years after it was completed. [37] I spoke with Kahn on three occasions about the response of the users. Each time Kahn admitted there were problems, indicated that he thought they had been exaggerated, and quickly turned the conversation to other topics. At the same time, we know he was hurt by the scientists' complaints. When he was designing the Salk Institute, Kahn brought his plans to the microbiology department, which was still housed in Richards. He hoped the plans for Salk would redeem his reputation, but the faculty was not appeased.

Major support for his morale during these years came from the tremendous enthusiasm for the architectural community for his work and ideas, including Richards. Scully's book about Kahn, identifying him as one of the masters of modern architecture, was published in 1962. It included several photographs and drawings reaffirming the brilliance of the Richards

Figure 5.6 The south elevation of Richards, photographed from the botanical garden during the early 1970s. The Biology Building is at the left, Tower D made up of studios is in the middle, and the Tower C, containing the animal quarters, is to the right. (Robert Gutman.)

parti and played down the functional problems. One of the illustrations included was of the studios "before partitioning." The true condition of the studio space is never shown in the book, although photographs of the completed building in use could easily have been obtained any time beginning in 1960. In 1961 the Museum of Modern Art in New York held an exhibition of the building, only the twelfth time in its history that the Department of Architecture and Design devoted an entire show to a single work of architecture. In an article about Richards, Wilder Green, who was curator of the exhibition, demonstrated his confidence in Kahn as an architect of great vision and substance by suggesting that one of the laboratories should have been left open "to demonstrate clearly Kahn's conception of the interior spaces."[38]

Kahn was aware of the complexity of the relations between the demands of architecture and the requirements of the modern building task. It was not easy for him, as it is not for any contemporary architect, to cope with the competing demands of the architectural tradition, clients, users, advanced building technology, contractors, and construction managers. I believe that a way out for him was the belief that his designs were the product of suprahuman and extraterrestial forces for which he was the spokesman and the draftsman. These convictions were the source of the well-known mystical quality in Kahn's architectural belief system. He indicated the forces to which he was responding operated below the level of rationally reflective consciousness. Kahn called the forces Psyche and Will.

> I think of Psyche as a kind of benevolence—not a single soul in each of us—but rather a prevalence from which each one of us always borrows a part. . . . And I feel that this Psyche is made of immeasurable aura. . . . I think that Psyche prevails over the entire universe. . . . I sense that the psychic Existence Will! calls on nature to make what it (Psyche) wants to be.[39]

Kahn thought that it was possible for an architect to discern the forms implied by these forces, if only he could leap beyond his immediate observations. He seems to have been confident that he had a gift for bridging the chasm between the world of sense perception and the more remote realms in which Psyche and Will operated. His confidence in his own abilities in this realm was a very central component of Kahn's architectural personality. It was this image of himself, as much as his awareness of his gifts as a designer and draftsman, that was the source of his authority in the architectural community. Architects in other firms, colleagues in his office, and students sensed something special about the man. Kahn was a truly charismatic figure, in the traditional meaning of the word—someone who was in touch with powers that transcended the sensibility and experience of ordinary persons.

As Kahn became older, and more successful, he became increasingly self-conscious about his special abilities. Like most relatively private persons who suddenly acquire a public persona, Kahn began to create a story about himself that would trace the origins of what an adoring community and he too defined as unique qualities. He told about his genealogical connections to old German rabbis and to Jewish mystics. Kahn talked at length about his mother, about her knowledge of German philosophy and literature, and the secret knowledge about the nature of the universe that she had inherited from her father, a rabbi, and that she intended to pass on to him. Kahn is said to have claimed that he was much influenced during the 1950s, the period when his philosophy was undergoing its most intense period of fermentation, by lectures on art and philosophy delivered by someone named Kunst, and later by conversations with Gabor, a Hungarian "philosopher" in Philadelphia who hung out at the Penn Department of Architecture and at Kahn's office. During the 1960s he has been reported to have been "reading" Jung's autobiography and the works of Schopenhauer.[40]

There has been very little writing so far on Kahn's career, his production, and his theoretical development that can be considered objective scholarship. Almost all of it is the work of former disciples and collaborators. The existing accounts will prove useful to the cultural historian, but for the present it is difficult to know what in Kahn's account of his intellectual journey is fact and how much is invention. In terms of the relation of his ideas to the tradition of philosophical and psychological thought, there *are* frequent examples of overlap, but exactly how Kahn absorbed the thought of, say, Schopenhauer or Jung, must for the moment remain a mystery.[41] It is clear, however, that, like Wright, he was more at home with German philosophy of the eighteenth and nineteenth centuries than with French thought, despite his architectural training under Paul Cret and the similarity between Kahn's compositional ideas and those of Choisy. In other words, Kahn's comments illustrate the electic use of philosophical repertoires that, at the beginning of this paper, I said was a characteristic of architectural thought and speech. The electicism, I wish to emphasize again, is consistent with the architect's natural propensity to express his ideas visually in the form of graphic images. The philosophy and the theories about human nature and motivation are adapted to support this objective. They also are used to explain the product of the visual exploration to an audience that is unfamiliar with the premises of architecture.

In the meantime, despite the lack of deep scholarship on the sources of Kahn's ideas and a more precise understanding of how he used them to arrive at his architectural solutions, we can conclude that Louis Kahn's architecture and thought as it developed in the 1950s and 1960s made a significant contribution in helping architects to overcome their dependence on a positivist model of man, in favor of a view that acknowledged the importance of the needs of the human spirit. Kahn was buttressed in the formulation of his philosophy by the conviction that his approach was exactly the perspective that the architectural community needed in the first decade after World War II. There was widespread dissatisfaction at the time with the program-driven functionalism of international modern architecture. Even the modifications in the CIAM platform introduced by Team X, whose American spokesman Kahn was, were inadequate to stem the rebellion against the architectural theories of Le Corbusier and Gropius and Gideion's historicist interpretation of the origins of modern architecture. The profession was searching for an architecture that could accommodate the new types of building construction and environmental control technology. But it was also looking for ideas that would enrich the human qualities of building space and that could connect building to some of the great formal ideas of the architectural tradition. Despite its practical failures, the Richards building exhibited a thrust that encapsulated these concerns. It also tied interest in these issues to theories of human nature and group needs that had been established in American architecture by the buildings of Sullivan and Wright. In harnessing this tradition that had been popular earlier in the twentieth century and rehabilitating an idea of functionalism that was more humanistic than the modern movement had suggested, Kahn offered a vision that was at the same time familiar and original. By making Richards into an emblem of his view of the nature of man and social institutions, Kahn became a cultural hero among architects and elevated a medical research laboratory into the architectural canon.

Notes

1. The trend toward a more autonomous architecture began in the 1950s, in part as a polemic against the anonymous spaces and buildings which were a by-product of the ideology of orthodox modernism. Richard Pommer lists several buildings that manifested this polemic, including the Richards building. They shared in common the aspiration to become "heroic works of art that would give meaning to the lives of their users." Richard Pommer, "The Art and Architecture Building at Yale, Once Again," *Burlington Magazine* 114 (1972): 860.

2. Argyris and Schon suggest that the unsystematic and eclectic approach to theoretical ideas is characteristic of all professions and practitioners and is a necessary feature of the world of action. They regard professional thought in architecture as just one more case of a more general difference between practical theorizing ("theory-in-use") and academic theorizing ("espoused theory"). Chris Argyris and Donald Schon, *Theory in Practice: Increasing Professional Effectiveness* (San Francisco: Jossey-Bass, 1974), especially Chapters 1 and 8.

3. Robert Venturi, *Complexity and Contradiction in Architecture* (New York: Museum of Modern Art, 1966), p. 20.

4. This is a paraphrase of statement made by Peter Eisenman in *The Charlottesville Tapes* (New York: Rizzoli, 1985), pp. 140–45.

5. In his fascinating account of his experiences with Kahn, the Pittsburgh architect and designer Wiliam Huff recalls a conversation on this theme:

   > Lou didn't care for the word "society." When I once grandly asserted that I wanted to "serve society," he rebuked, "Society doesn't want to be served. Society doesn't care." *Society* wasn't a word for him; institution was. Institution has the aspiration of *man*, not men, man. He thought the great thing that an architect could do was to help, in his limited way, the institution be something better—to realize its original aspirations."

   William Huff, "Louis Kahn: Sorted Recollections and Lapses in Familiarities," *Little Journal*, 5 (September 1981): 7 (publication of the Western New York Chapter of the Society of Architectural Historians).

6. Peter Smithson, "Thinking of Louis I. Kahn," *Louis I. Kahn* (Tokyo: A + U Publishing Co., 1975), p. 321.

7. The process of documenting the building's problems is made easier because the unconventional character of the design generated a torrent of criticism that produced an adversary situation early during the design and review process. This criticism continued to be manifested after the building was finished and the first scientists moved in and persists to the present day. The intensity of feeling, in turn, brought forth a flood of memoranda, letters, studies, and reports both from the the university side and from the architect.

   Much of the material not published in books that is quoted in this paper was obtained by examining the files at the Kahn Collection located in the Furness Library of the Graduate School of Fine Arts at the University of Pennsylvania. I am grateful to Ms. Julia Converse, Curator of the Collection, for her guidance and advice about the materials available in the file. Most of this material is still uncatalogued.

The other unpublished material used in this study was gathered from interviews I conducted personally during frequent visits to the Richards Labs, beginning in 1970. In addition, some of the major participants in the programming and design process during the period 1957–58 were good enough to allow me to see some of their correspondence with Kahn and with the university administration.

I have reported other findings from this research in three papers published earlier. "The Evaluation of the Constructed Environment: A Comparison of Two Bio-Medical Research Laboratories," in *The Constructed Environment with Man as a Measure*, ed. Edwin A. AbdunNur (New York: American Society of Civil Engineers, 1976), pp. 14–46; "Building Evaluation, User Satisfaction and Design" (with Barbara Westergaard) in *Designing for Human Behavior*, ed. John Lang et al. (Philadelphia: Dowden, Hutchinson and Ross, 1974), pp. 320–29; "Who Decides What a Building Wants to Be?: A Study of Louis Kahn and His Clients," Association of the College Schools of Architecture, *Proceedings of the Annual Convention, Vancouver, B.C., 1985* (Washington, D.C.: ACSA, 1986).

8. Kahn participated in a meeting at Yale in 1953 at which Johnson Wax was discussed, although he did not visit the building until 1959. At the meeting, which included Philip Johnson and Pietro Belluschi, the practical problems of the Tower were mentioned frequently. Kahn's comments were ambiguous and revealed his ambivalence but offer a clue to how he might have felt about the criticisms of Richards when *his* lab tower had been built. Here is what he said:

> The Tower was done with love and I should say it is architecture. It belongs to Mr. Wright personally. It belongs not to the sociological aspect of architecture so much as to the physics book of architecture. . . . If the Tower has this power to throw out sparks, to make you want to build one of these things, then I believe it functions. If it doesn't necessarily function as an experimental laboratory, then Wright should be fired by the Johnson Co. The form itself does excite us.

"On the Responsibility of the Architect," *Perspecta* 3 (1955): 47.

9. The design of the service towers may have been influenced by Eero Saarinen's design for the General Motors Technical Center of 1949–55 in Warren, Michigan. In this building the mechanical system is clearly articulated by the oversized exhaust shafts that are vertical elements on the building's exterior.

In Richards many of the services are not located where the theory of the building says they should be. They run up and down through the central core, near the elevators and the laboratories where the animals used in experimental research are housed. It proved too cumbersome and expensive to send all the utilities through the servant towers.

10. John Cook and Heinrich Klotz, *Conversations with Architects* (New York: Praeger, 1973), p. 212.

11. From Kahn's lecture in the Voice of America Forum series, 1960, quoted in Vincent Scully, *Louis I. Kahn.* (New York: Braziller, 1962), p. 120.

12. Seven to eight years after Richards was completed, a small project was built in the Old Medical School building that incorporated some of the same ideas about the community of science that Kahn used to justify the open studio. In this case, walls and partitions that had been constructed before the turn of the century were *taken down*. They were replaced by a lattice work of shelves designed to express and encourage interactions between members of established medical disciplines. The project was for a research program in immunology. The renovations were designed by the scientists themselves because they assumed there was no architect who could appreciate their need for a built environment that would encourage interchange between disciplines. The young immu-

nologists might have been the ideal tenants for the original Richards building, or for a new facility designed by Kahn. However, by the middle of the 1960s the reputation of the Richards building as a research building was sufficiently ignominious that only low-status groups in the medical school could be forced to move into it. And Kahn, by this date, was no longer on the list of approved architects at Penn.

13. University of Pennsylvania Project Files: Memo of Meetings. Kahn Collection.
14. University of Pennsylvania Files: Medical Research Center—Data from Doctors. Kahn Collection.
15. University of Pennsylvania Project. Job Meeting Files. Kahn Collection.
16. Letter to the author, January 8, 1971.
17. Joseph Burton, "Notes from Volume Zero: Louis Kahn and the Language of God," *Perspecta* 20 (1983): 85.
18. The engineer Kommendant reports a similar impression. See August E. Kommendant, *18 Years with Architect Louis I. Kahn* (Englewood, N.J.: Aloray, 1975), p. 20.
19. University of Pennsylvania Project. Job Meeting Files. Kahn Collection.
20. For example, a plaintive letter from the dean of the medical school sent in January 1959 to the heads of departments going into Richards pointed out that their estimates for *movable* equipment alone totaled $1.1 million. The university had $260,000 available, one half of which was federal funds.
21. Beginning in 1985, however, I noticed a change in sentiment about Kahn among those in charge of Richards, and a certain respect for him and the building, too. In part, this is because some of the scientific disciplines that found the building most awkward for their work have been moved to other facilities around the medical school. Also, as more space has been built for the medical school, the faculty and research staff is less dependent on Richards, and so the building can be revered for its architecture. One sign of the new respect for the building is that Marshall Meyer, Kahn's principal design collaborator in the last projects of his career, is doing the latest series of renovations to the building.
22. Alexandra Tyng, *Beginnings: Louis I. Kahn's Philosophy of Architecture* (New York: Wiley, 1984), p. 79.
23. Huff, "Louis Kahn: Sorted Recollections and Lapses in Familiarities," p. 7.
24. William Jordy, "Medical Research Building for the University of Pennsylvania," *Architectural Review* 129 (1961): 100.
25. *Architecture and the University: Proceedings of a Conference held at Princeton University, December 11 and 12, 1953* (Princeton, N.J.: School of Architecture, Princeton University, 1954), pp. 29–30, 67–68; and Alexandra Tyng, *Beginning: Louis I. Kahn's Philosophy of Architecture* (New York: Wiley, 1984), pp. 63–64.
26. Norris Kelly Smith, *Frank Lloyd Wright: A Study in Architectural Content* (Englewood Cliffs, N.J.: Prentice-Hall, 1966), p. 41.
27. Joseph Burton, "Notes from Volume Zero: Louis Kahn and the Language of God," *Perspecta* 20 (1983): 71.
28. The concept of form is similar to the idea of building types that became the fundamental theory of architectural design during the 1970s. The term *form* has a Germanic root, whereas the idea of *type* is originally French. For an illustration of the role of typology in contemporary architectural theory, see Alan Colquohoun, "Typology and Design Method," reprinted in Robert Gutman, *People and Buildings* (New York: Basic Books, 1972).
29. Louis I. Kahn, *Talks with Student: Architecture at Rice* (Houston: Rice University, 1969), pp. 7–8.
30. "Panel on Philosophical Horizons," *American Institute of Architects Journal* 33 (June 1960): 99.
31. Kommendant, *18 Years with Architect Louis I. Kahn*, pp. 7–8.
32. Smith, *Frank Lloyd Wright*, pp. 140–45.

33. "Panel on Philosophical Horizons," p. 100.
34. Cook and Klotz, *Conversations with Architects*, pp. 192–93.
35. Scully, *Louis I. Kahn*, p. 30.
36. These later events are reported in my paper, "Ethical Issues in the Building Process," *Via*, forthcoming.
37. For a typical instance of Kahn's statements about the building, see his Voice of America Forum Lecture, in 1960, quoted in Scully, *Louis I. Kahn*, p. 119–20.
38. Wilder Green, "Medical Research Buildings—Louis Kahn," *Arts and Architecture* 78 (1961): 17.
39. Burton, "Notes from Volume Zero," p. 70.
40. It would seem from the context in which Tyng mentions Kunst that he may have been lecturing at Yale. See Tyng, *Beginnings*, p. 17. Kahn described Gabor as a "man in my office who doesn't do any work. But I gladly pay him because he helps me think." See Tyng, *Beginnings*, p. 168. Kahn's reading of Jung and Schopenhauer are mentioned in the article by Burton and the book by Kommendant, both cited previously.
41. To take just one example of the similarity between the attitudes toward architecture of Kahn and Schopenhauer, there is the following sentence: "architecture does not affect us mathematically, but also dynamically . . . what speaks to us through it is not mere form and symmetry, but rather those fundamental forces of nature, those first Ideas, those lowest grades of the objectivity of will." The passage is from Schopenhauer's major work, *The World As Will and Idea*.

Robert Kerr:
Architect of Bearwood

KENT BLOOMER

We may refer to the Victorians as stiff in their social manners and at the same time regard their architecture as wildly romantic, with its protrusions, polychromy, and proliferation of elements. Those two impressions, placed together, reveal apparently contradictory expressions of a new industrial class in England that was anxious, on the one hand, to behave properly and to confirm its ascent politely, and that, on the other hand, inclined toward exuberant celebration.

Most of the major English architectural theoreticians of the nineteenth century manifested those "psychological" extremes in their texts. A. W. Pugin, in his *True Principles of Pointed or Christian Architecture*, first published in 1841, demanded a dutiful integration of engineering details as prominent elements in the design of modern architecture while at the same time praising the divine power of spires and proposing a return to the decorated Gothic of the thirteenth and fourteenth centuries. John Ruskin, in *The Seven Lamps of Architecture* (1848), insisted that materials be "truthful" expressions of their own intrinsic nature, although they were expected to serve other purposes as well, particularly the requirement that buildings, in their whole fabric, represent and even mimic the savage and beautiful forms of the natural world outside. Owen Jones's monumental *Grammar of Ornament*, written in 1856, introduced twenty-one "scientific" principles governing the harmonic use of colors in order to achieve "repose." However, his rational principles, carried into his own work, produced designs so vibrant that even today we would most likely describe them as psychedelic rather than harmonic.

Robert Kerr, one of the most articulate architectural journalists of nineteenth-century Britain and a founder of the present Architectural Association in London, seemed in his lectures and writings to be supremely rational in his approach to the social planning and zoning of rooms. In 1864 he wrote a book called *The Gentleman's House* that immediately attracted a commission for his design of an enormous country house called Bearwood, which was destined to seal in brick and stone his concept of people. Bearwood reflected Kerr's vision of a community benevolently dominated by a single person, the Victorian Gentleman, whom he believed had evolved over one thousand years from a rude Saxon thane to the proprietor of a highly ordered rural estate in which everyone (in contrast to the primitive Saxon household when the members huddled together in a single drafty and muddy hall) would have carefully articulated private rooms and workspaces, all separate and identified but very much subject to the rational organization of daily Victorian life.

The Gentleman's House

A history of that evolution, as well as Kerr's belief that it was evidence of benevolent progress, is outlined in "Part First" of his book. He begins his history with the eleventh-century Saxon hall, rather than with the earlier Roman houses that once occupied Britain, because of its indigenous northern character. He characterizes a single covered enclosure with an open hearth as the principal space of England's truly native architecture.

> The ordinary Saxon Hall constituted the sole dwelling-room and eating room, for lord and lady, guest and serf . . . kitchen and scullery . . . and quarters none the less for the sheepdogs and wolfhounds. It afforded stowage in one corner for implements of husbandry, and in another for a store of produce. Lastly it was the one universal sleeping room of the household, who disposed themselves according to their rank upon the floor.[1]

Indigenous and northern, perhaps, but also rude and disordered for its lack of privacy and inarticulate work and storage space. From that primitive beginning the careful division, amplification, and specification of space could only, by Kerr's analysis, be for the better.

The earliest divisions were between the main hall, a cellar for beer and animals, a kitchen, and an attached chamber for important business transactions away from the common company. In royal houses a second principal chamber, or chapel, was attached to the main hall for a place to worship as well as to conduct the affairs of state.

It was not until the thirteenth century that the private bedchamber was introduced by architecturally dividing one of the attached "chambers" into a bedchamber and parlor. In that development a wall replaced the lighter screens or wood partitions formerly used to segregate sleeping space from sitting space of daytime for nighttime functions. Kerr notes that the privilege of privacy was afforded by rank.

> The persons of chief importance in a noble household were three, the lord, the lady, and the priest. . . . The pride of the lord was not yet of that kind which we call exclusiveness; the fastidiousness of the lady was all undeveloped; the contemplative occupation of the priest demanded quiet. Accordingly, beside the chapel we have seen the priest's chamber; and this constituted, we must say, the first properly private apartment in an Englishman's house.[2]

In due course the lord was to acquire privacy, and fairly late in the evolution of domestic space the lady, originally by the division of a Queen's chamber, acquired a private bedroom.

By the fourteenth century the original hall was still the architectural center of the house. Over time that hall had become enormous and elaborate on a par with ecclesiastical architecture, despite its use as a space for common feasts, after which the guests of both sexes would sleep on the floor. Indeed, the public clutter of this central gathering place prompted royal families to specify a private dining room and set aside a bower or loft in the chilly attic as a dormitory for servants.

Apparently the slow departure of the lord and lady from the center of the house signaled the eventual decline of the great hall as the principal ceremonial place, and as a consequence, during the fifteenth century, it became reduced in relative size as new kinds of rooms proliferated throughout the house. Within those rooms, which included pantry, scullery, and bakehouse as well as additional offices for the management of large estates, cupboards and closets were invented to replace the traditional portable chest. Specific closet types, such as wardrobes or washing closets, also began to prescribe special functions to rooms.

This process of particularization by functional subdivision increased throughout the sixteenth century, so that in great manor houses eating rooms were split between breakfast rooms and dining rooms, parlors proliferated into summer and winter parlors, more private bedrooms identified members of the household, an "inferior hall" for servants appeared near the kitchen, and the boudoir was imported from France as a separate sitting or dressing room for women. To accommodate clusters of rooms the corridor was created, which, generations later, when sufficiently enlarged to marshal large gatherings for public rooms, ushered in the supercorridor, which was capable of exhibiting portraits and providing seating for visitors. This new room, sometimes called the gallery, emerged to displace the ancient hall further as a principal public room.

Kerr was clearly impressed by the invention and distinction of new room types as exemplary of an architecture responding to the complex functions

Figure 6.1 Rufford Hall, Lancashire, England.
Over time the Hall had become enormous and elaborate on a par with ecclesiastical architecture despite its use as a space for common feasts after which the guests of both sexes would sleep on the floor. (Chadwyck-Healey Ltd.)

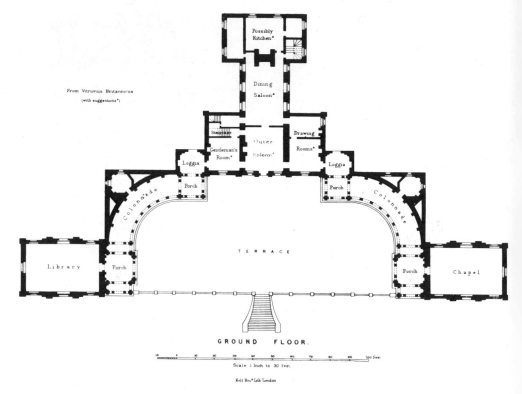

Figure 6.2 Stoke Hall, Northhamptonshire, England.
Stoke Park, a country house by Inigo Jones, had the library and chapel situated in wingrooms, which were equal in shape and disposition. (Chadwyck-Healey Ltd.)

of domestic life, particularly English domestic life. He associated room specialization with the development of comfort, privacy, and convenience. He
equated civility with an efficiency gained from properly identifying, articulating, and arranging emerging differences that, in primitive houses, were
denied expression.

When he records that this progress was seriously threatened during the
seventeenth century by the importation into England of the Italian villa
type by Inigo-Jones, who had visited and mastered Palladio's work while
living in Italy, he despaired. He believed that the alien Mediterranean classicism imposed an unnatural symmetry and a false geometry on the arrangement of rooms that would require that different functions be housed
in identical rooms. Stoke Park, a country house by Inigo-Jones, had the
library and chapel housed in wingrooms that were equal in shape and disposition.

In Marlborough House, designed in 1709 by Christopher Wren, the

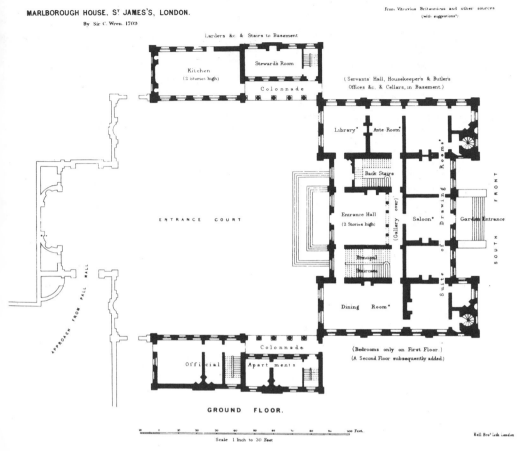

MARLBOROUGH HOUSE, S^T JAMES'S, LONDON.
By Sir C. Wren. 1709

From Vitruvius Britannicus and other sources
(with suggestions')

GROUND   FLOOR.

Scale 1 Inch to 30 Feet

Figure 6.3 Marlborough House, St. James's, London.
In Marlborough House, designed in 1709 by Christopher Wren, the kitchen faced forward
on the left wing and the official apartments on the right with identical ornaments.
(Chadwyck-Healey Ltd.)

kitchen faced forward on the left wing and the official apartments on the
right with identical ornaments and dimensions on their respective façades,
and the principal staircase and back staircase were given equal status for
having equivalent access in plan on either side of the central entrance hall.

Kerr declared that such arrangements represented an urban pretension
"charged with seeming more like some temple for the gods rather than the
[country] home of an English family."[3] He did not object to the invention
and importation of new room types imported from eighteenth-century neo-
classical houses such as the ballroom, the billiard room, and the music
room; but he deplored their subjection to a regulating geometry that did

not originate, so he implied, in human behavior. Beauty, which he acknowledged was a public presence or at least an "effect" in Anglo-classicism, would, if allowed to determine form, favor both inconvenience and extravagance.

By the nineteenth century Kerr felt that although a complete inventory of rooms had come of age that was quite capable of providing the privacy and convenience unavailable in the ancient halls and early manor houses, it was often denied expression by a fashion that crippled the proper arrangement of rooms with wasteful porticoes, wings, arbitrary symmetry, and the excessive publicness of an interior cortile or covered Mediterranean courtyard. He suggested more than once that those devices belonged to another age because the Englishman had become sufficiently refined to be the subject of his own place and did not need to posture as a divinity from antiquity.

Despite his consistent references to great mansions and royal abodes, Robert Kerr actually intended the architecture of "the gentleman's house" to be available to a large constituency that might reflect the domestic habits of refined persons, although he states quite clearly "it is not meant to represent inferior dwellings such as cottages, farmhouses, and houses of business."[4] He does allow that "persons who have been accustomed to the best society find themselves at ease; and there are others upon which ample dimensions, liberal outlay, and elaborated decoration have entirely failed to confer the character of a Gentleman's House."[5] In any case, Kerr is consistent in his method of treating good architectural form as that which emanates from the top down by showing that new room types and arrangements, such as woman's privacy, were created in royal abodes first and that subsequently places for privacy, such as bedrooms, were ultimately to become everyone's property.

He states at least one outstanding principle in his summation:

> Primarily the House of an English gentleman is divisible into two departments; namely that of THE FAMILY, and that of THE SERVANTS. In dwellings of inferior class, such as Farmhouses and the houses of tradesmen, this separation is not so distinct; but in the smallest establishment of the kind with which we have here to deal this element of character must be considered essential; and as the importance of the family increase the distinction is widened—each department becoming more and more amplified and elaborated in a direction contrary to that of the other.[6]

Bearwood

What, then, did Robert Kerr design when awarded the spectacular commission shortly after completing his book? His client was John Walter, chief

Figure 6.4 Bearwood, Berkshire, England.
The house is greater than 300 feet in length, 88 feet to tower top, and contains more
than 100 rooms. (Chadwyck-Healey Ltd.)

proprietor of the *London Times* and wealthy member of the new Victorian
professional class that acquired power during the industrial revolution.
"Neither John Walter nor his family ever seems to have shown much am-
bition to mix or marry into the aristocracy."[7] The program for a new and
truly modern country seat was to house a large family on a recently developed property in Berkshire County. It appears that Kerr had a free hand in
design along with a budget that enabled him to articulate his theories in a
single stroke.

The house is greater than 300 feet in length, 88 feet to tower top, and
contains more than 100 rooms. It is approached by a straight driveway that
rises gradually between rows of California redwood, stands on the crest of
a hill, and is terraced behind down to a lake surrounded by evergreen woods.
The most prominent feature upon arrival is the immense tower to the left
of the main entrance, establishing a center of gravity for the wide composition of the façade. It also divides the department of family on the right
from that of servants on the left. Immediately Kerr's doctrine is visible in
the massing, the family being provided with an architecturally separate entity in the form of an enormous and somewhat conventional "three-story"
house centered about a main entrance that, without the main tower, could
probably be reduced to a fifth of its volume and be perceived as a typical
"villa."

The interior plan, however, belies the typical villa plan by carrying to an

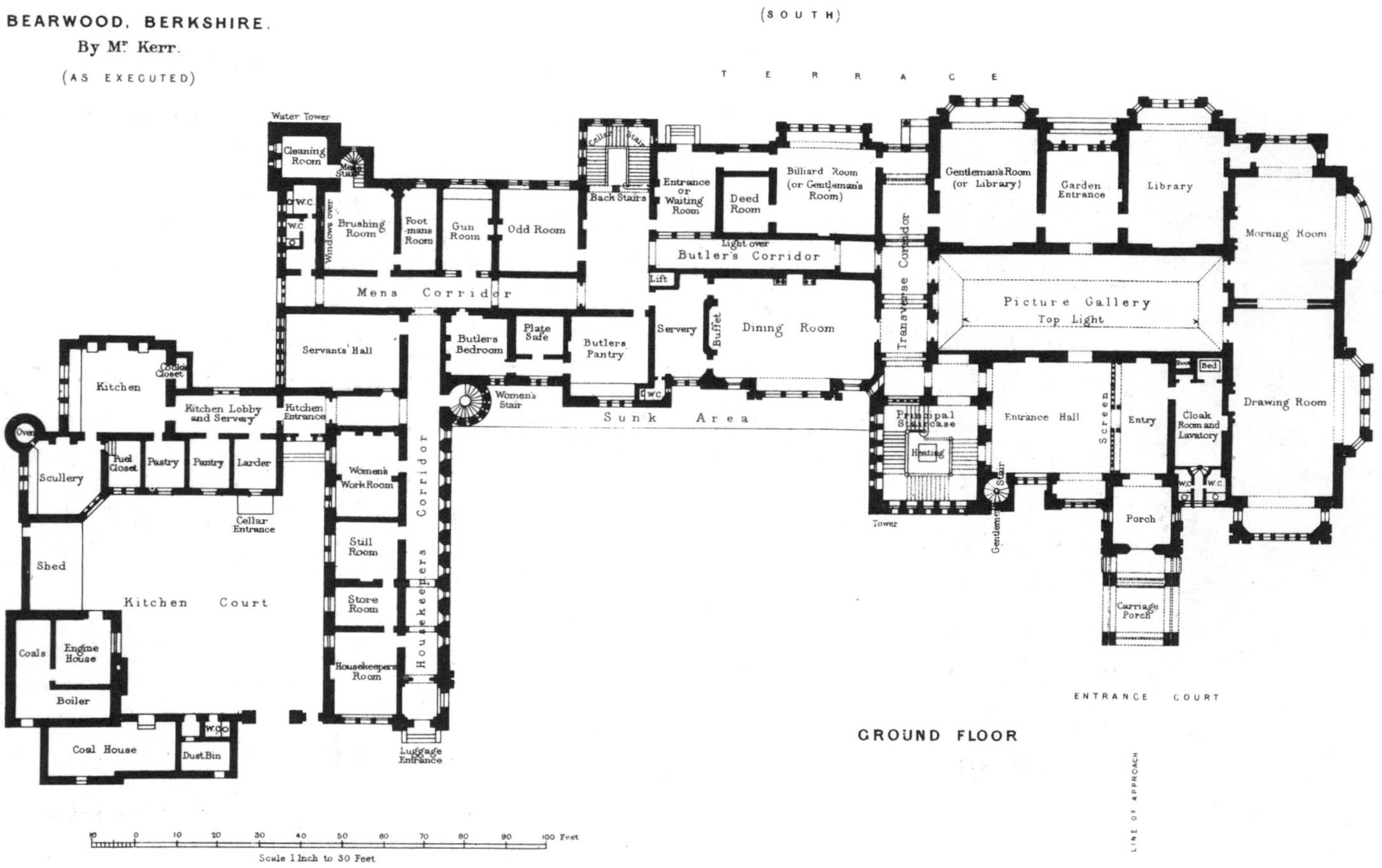

Figure 6.5 Plan of Bearwood.

The interior plan belies the typical villa plan by carrying to an extreme Kerr's doctrines of subdivision and differentiation. (Chadwyck-Healey Ltd.)

extreme Kerr's doctrines of subdivision and differentiation. One enters through a carriage porch into an enclosed porch and thence into an evocation of the Saxon Hall complete with a screen, a vestige of the ancient divider between door and central hearth that functioned originally to keep wind away from the fire.[8] The hall provides access to the principal staircase as well as the principal corridor, which in Bearwood is a skylit picture gallery planned for Mr. Walter's collection of Flemish paintings. From that gallery, all other corridors are accessible, and they include a transverse corridor, a butler's corridor, a men's corridor, a housekeeper's corridor, and upstairs, a family corridor, a nursery corridor, and a women servants' corridor.

The proliferation of stairway types is even more astounding. There is a principal staircase, a gentlemen's stair, a bachelor's stair, a young ladies' stair, backstairs, a women's stair and a strangers' men servants' stair (servants of guests). Curiously there is no "older" ladies' stair. The women's stair is meant only for women servants, despite the spirit of Kerr's special concern in his chapter on stairs that

> the principal staircase ought to be so placed as to afford direct passage, for the ladies particularly, from the public rooms to the bedrooms; and secondly access from entrance ought to be equally direct, for the ladies again, when coming from out of doors, so that they may not have to pass through any great extent of interior thoroughfare.[9]

Actually, the men are better provided to walk upstairs out of sight either by using the gentlemen's stairs off the main entrance hall or coming in the rear entrance waiting room and going up the backstairs. Even the young ladies, in contrast to the bachelors, do not have their own passageway from the ground floor to the bedrooms, although they are provided with one adjacent to the ladies suite upstairs for access to the third level. It might be unfair to fault Robert Kerr for overtly favoring men in these instances because he probably imagined himself to be a proponent of women's privacy by speaking out for their needs and at least providing a novel stairway for young ladies in 1864.

Nevertheless, the advantage for men is evident again when we examine the entrances to the house in relation to public rooms and special rooms. The main entrance and garden entrance are clearly intended for both ladies and gentlemen, but the back entrance seems off bounds for ladies, because in addition to providing access to the servants' wing, it only provides a direct exterior access to the deed room and billiard (or gentlemen's) room. Men could easily enter the house from behind, shoot some billiards, collect their guns, and then go off to the woods virtually without being seen or appearing in the public rooms. Appropriately, the nearest public room to the billiard room is the library, which is also designated a gentlemen's room. Ladies, on the other hand, must pass through the principal corridors and staircase before arrival to their one completely private suite, which is a connected boudoir and ladies' drawing room upstairs on the garden side.

The sociological order is even more pronounced within the servants' department. A single common space, the Servants' Hall, is provided with a fireplace and large window for viewing outward, which minimally conforms to Kerr's advice that the servants' principal prospect not be disagreeable (the prospect is interrupted by a kitchen wall), but fully conforms with his proviso that "the outlook, however, ought not to be towards the walks of the family; neither need it be towards the approach."[10] Indeed, all the servants' bedrooms and offices are denied views of the garden and only the lady's maids and butler hold sufficient rank to warrant a view from their bedrooms toward the approach. The housekeeper, as the highest-ranking woman domestic, is given a view of the approach from her office but not from her bedroom.

Women servants, on the scale of Kerr's social evolution, are held slightly behind the men servants, particularly in matters of privacy.

> The ordinary female domestics are usually provided with bedrooms on the uppermost story, or over the offices, accessible by the backstaircase. These rooms ought to be of small size, suitable for not more than two persons. They ought as a whole to be grouped together. Every room ought to have a fireplace, and good light and ventilation. The ordinary men-servants must have their sleeping rooms in a separate quarter. Each man ought properly to have a separate room. . . . The upper servants of each sex will expect to have separate rooms as follows. The housekeeper ought to sleep near the maid servants; the lady's maid if possible near her mistress, a woman cook may have a separate room amongst the others; a man cook will have a room near the kitchen; the butler will sleep near the pantry; and if the valet can be put within reach of his master during the day, so much the better for the efficiency of his attendance.[11]

With the exception of the Servants' Hall the only "sitting" rooms provided for the servants at Bearwood seem to be ad hoc. Possession of those rooms is prescribed by Kerr in *The Gentleman's House:*

> Throughout the whole a sort of principle is held to govern, that any sitting-room of the servants is only conditionally private, and may be open, according to arrangement, to some partner of the same rank; and this more particularly when the accommodation of visitor's attendants comes into question.[12]

Kerr believed that hospitality should provide the guests' servants with accommodations slightly superior to those of the domestics and even specifies a room in Bearwood as a Stranger's Nursery or sick room.

Robert Kerr's People

Certainly Robert Kerr specified the degrees of authority, privacy, territory, and amenity that he allowed members of a large English Victorian household and was fastidious in the translation of his beliefs into the planning of Bearwood. He was also forthright in stating that it was a gentleman's house, which we may take quite literally to mean a house that represents a man rather than a woman, or a divinity, as well as a house that does not, in Ruskinian terms, primarily represent a landscape or a work of craft and technology. Mark Girouard suggests that Kerr attempted to make the design of the façade "muscular . . . the entrance front is like a sock on the jaw," and unquestionably the impression upon arriving at the house is one of uncompromising authority.[13]

However, Kerr also states that

the qualities which an English gentleman of the present day values in his house are comprehensively these: Quiet comfort for his family and guests; Thorough convenience for his domestics, [and] Elegance and importance without ostentation.[14]

He further states that

the account which has been given of the history of the Plan will pretty clearly show in what manner and by what degrees these principles have come to be established, and how recently it is that they have been fully recognized; but it is none the less certain that at the present moment they must be considered to be *fixed and final conditions*, of which no compromise ought to be proposed by a competent designer.[15]

The notion of "fixed and final" seems an anomaly in the framework of Kerr's panorama of history, which is so dependent on a process of evolution. Indeed, the status of his client, John Walter, was itself a product of social mobility for having benefited from the power unleashed by the invention of the steam-driven printing press and the industrial revolution.

Perhaps Kerr's people are really his contrivance of an eternal English person who hadn't "evolved" as much as having been unearthed by the phenomena of subdivision, particularization, and amplification. In that respect they were always in the woodwork, so to say, but denied explication. It seems to me that Kerr imagined them to be fundamentally pleasant and accommodating people who were strictly disciplined, ranked, and dependent on one another, although his rank-ordering allowed few opportunities for most members of the Bearwood community to express themselves.

The disappointing property of the architecture of Bearwood resides paradoxically in its quality of homogeneity of architectural sameness. That feature was the most unexpected during my visit to the house, especially in light of Kerr's promise that different social distinctions were to be represented in the architecture of the rooms, corridors, and staircases. How, other than location, does the men servants' stair differ in feeling from the young ladies' stair, or how do the corridors in the nursery differ basically from those assigned to the domain of the butler or men servants, or how does the form of the partitioned "ancient" Hall call out memories that differ from the Library of the nineteenth century? While the distinctions appear in the organization, reinforced on paper with words, the reality is one of oversized yet architecturally similar spaces ornamented in a thin pastiche typical of the Victorian fashions of the 1860s and serviced by corridors so wide, long, and institutional that they are more reminiscent of a 1930s American high school or office corridors meant for heavy traffic than those belonging to a family residence, even if that residence is a country seat.

The central corridor gallery fails to deliver either the impact of a supercorridor or the centerplace of a cortile, but instead appears to be a bland museum without particular distinction. It is difficult to imagine breakfasting in the morning room, one of Kerr's specified functions, not because it is 300 feet from the kitchen, but because it is so big and so like the shape and height of the libraries, drawing room, and the dining room, directions of sunlight notwithstanding, that it lacks specialness.

Robert Kerr was able to deliver the organizational diagram that differentiated his people but not a profoundly differentiated work of architecture; and in so doing presaged some of the anonymity characteristic of the more mechanistic and functionally planned buildings we have come to deplore in the late twentieth century. The rational properties of Kerr's analysis of history, translated into a flow diagram, are more pervasive than the ceremony, however forbidding, of the ancient manor houses that he claims are so innate to his vision.

It was particularly disappointing to visit the special secondary stairs. In searching for words to best describe their quality it seems enough to say that they were like well-constructed internal fire escapes rather than social events.

The principal staircase and tower, on the other hand, is a spectacular and brooding achievement. That entity, which is diabolically intertwined with the house, is so fantastic that, with a single gesture, it almost makes up for the peculiar ordinariness of the whole, both inside and out. Primarily as a marker for the entire ensemble, anchoring the house to the courtyard, it intersects and aborts the symmetry of the left-hand to the right-hand gable of the entrance façade. Conversely, the left-hand gable crashes into the tower as though it were the head of an axe in the side of a tree. To compensate for the asymmetry above, the first tier of windows in the base of the tower rises upward to the right, enshrined in belt-courses that establish a larger corner-buttress to the right than to the left of the base. Nestled in the midst of the collision is the turret for the Gentlemen's and Bachelor's stairs, the form of which is reiterated on top and in back of the tower as an oriel reaching upward to the highest point and topped off with a finial holding arrows pointing north, south, east, and west. That magnificent, pagan, and irrational tower is the finest architectural moment and may very well be the icon and genius loci of Bearwood.

Robert Kerr's people might thus be characterized by initially dismissing the notion that they are either traditional characters or altogether "real" people. In his text he campaigns against the classical house tradition, which he claims is more appropriate for obsolete gods than for English families. Yet, when he designs for his laboriously registered members of the English household, he treats them impassionately by casting their architectural expression predominantly within a diagrammatic as well as a sociological concept that he admits is frozen. Although he expresses an abiding concern for those people, his animation of them falls short.

But neither Kerr nor the spirit of his tower evidence an incapacity for passion. Although the principal staircase is designated for the family as a whole, it is more likely that the tower is the architectural expression of an industrial demiurge that, rather than being frozen in place, is marvelously energized and aspiring, and within which one ascends helically, underneath a star-studded ceiling, and beside enormously tall windows, with a sensation of emergence more than convergence into a fixed and final place.

It is difficult to determine whether the demiurge is John Walter or Robert Kerr, but that does not really matter. As Kerr stated, the house was the property of an *individual*. In the circumstances of Bearwood that individual was cast as a captain of industry commanding a domestic institution remarkably prophetic of a twentieth-century corporate structure. The people in that corporation, the people of Bearwood, are mechanistically defined and expected to behave like the gears and spindles of John Walter's printing press.

The internal workings of the machine, characterized by Kerr as benevolent and harmonious, represent his "polite" and rational side, whereas the tower reveals his aggressive and exuberant side. The social structure of the household is prescribed and frozen, whereas the "genius loci" is mobile and emotional. Can we believe that Robert Kerr did not suspect that the very individualism he granted John Walter was not on the way to becoming public property? Did he not understand that he was describing a process of

Figure 6.6 The Tower of Bearwood.
Neither Kerr nor the spirit of his tower evidence an incapacity for passion. (Chadwyck-Healey Ltd.)

evolution rather than of prescription? Did he not read Charles Darwin's *Origin of the Species* published five years before *The Gentleman's House?* Or if he did, was he rejecting Darwin's thesis? Whichever, he was the true Victorian, mixing stiff social manners with wildly romantic protuberance.

Notes

1. Robert Kerr, *The Gentleman's House* (New York: Johnson Reprint Corp., 1972; originally published 1864), p. 4.
2. Ibid., p. 27.
3. Ibid., p. 43.
4. Ibid., p. 63.
5. Ibid.
6. Ibid.
7. Girouard Mark, *The Victorian Country House*, (New Haven: Yale University Press, 1979), p. 263.
8. Olive Cook, *The English Country House*, (London: Thames and Hudson, 1974), p. 12.
9. Kerr, *The Gentleman's House*, p. 176.
10. Ibid., p. 231.
11. Ibid., p. 250.
12. Ibid., p. 252.
13. Girouard, *The Victorian Country House*, p. 269.
14. Kerr, *The Gentleman's House*, p. 66.
15. Ibid.

7

The Master of the House

LARS LERUP

The Western house is the empire of the dweller, who, inscribed in the form of geometricized marks (such as chairs, sofas, tables, beds, windows, and doors), dominates the shape and layout of the house. This representation of the dweller or subject is a particular obsession of Western culture that seems to bar other potential relationships between subject and house. In fact, it is this obsession that prohibits the architecture of the house from becoming anything but an imprint of a stylized rendering of the physiological and social dimensions of the subject. Even if we accept the architectural function of the typical single-family house to be solely the representation of the subject, it seems clear that the fixed smile of the three-bedroom house is becoming dumber by the day.

The Single-Family House is what Durkheimian sociologists would call a "collective representation." Although it is clearly a construct, through usage, acceptance, aspiration, and commitment, it has become something we take for granted—in Marx's famous dictum: The single-family house has been "inverted" into an object of nature. The magnitude of this commitment is demonstrated by the fact that we now accept the single-family house as the token and locus of the "good life" and as our overwhelming housing preference, not just in America, but in places as disparate as Sweden and Cuba. The institution of the single-family house cuts across all cultural, geographic, economic, and political differences to form the mythical abode of the modern family.

To denounce the single-family house is not only futile, but empty posturing, unless shifted away from the surface of the myth to its interior, to its underpinnings, to the moving parts that suspend its meaning. The issue here is not to save the world from "the evil empire" of the single-family house, but to break it open, to reveal its artifice. This assault leads to a clear partitioning of the idea of the house from its form and to a suggestion that the idea itself needs revitalization and expansion.

The Master of the House

The dweller rules the design of the single-family house. More specifically, a fictitious dweller is the subject of the house.

Entering from the garage, we find the door itself is a Cartesian inscription of a person. The living room beyond, with its sofa, coffee table, and two overstuffed chairs suggests an endless kaffe klatsch, interrupted only by the opaque eye of the silent TV set obliquely placed so both sofa and chair can "see" it. The dining room, with six chairs around the dining table, combined with four seats around the coffee table, begins to outline the calculus of social interaction: two absent or present guests, four adults, two children, and so on. The largest bedroom with its master bed is the central inscription of intimacy in the house, followed hierarchically by a smaller bedroom with two beds (for the two girls), which in turn is followed by the solitary bed in the other bedroom, equal in size (for the son, who will go to law school). Returning by way of the corridor that connects the bedrooms, and noting the difference between the two bathrooms with their almost physical imprints of the complete body or its parts, the dweller returns to the garage, where the family car, by its sheer size, seems to be the only other contender of propriety in the house.

The single-family house serves as the frame of a stylized image of the family and its relationships. Like a picture frame, its role as mere support is underestimated. Having left the house, reflective dwellers know that their "fictitious consciousness" as inscribed is limited to the most mundane doings of everyday life. The full "inscription" of the dweller is not only the imprints but a complete economy of relationships that are played out on the map of the house, both as actual and implied behavior (advertising, home magazines, TV soap operas, etc.). The single-family house is a complex of inscriptions, behaviors, texts, and images, which assembled form the substance and ideology of the suburban house—in Foucault's words, a "disciplinary mechanism."[1] Against the tableau of this single-family house, we will place a conception of another house.

An actual house is only a model of the concept or idea of a house. This idea holds more or less explicitly the collective memory of all the houses we have conceived, built, and lived in. It is a complex one, with cultural, social, economic, ideological, aesthetic, and physical dimensions. Obviously, we can expect to comprehend only a fragment of this vast "book" at any one time. The single-family house occupies a section of this book.[2]

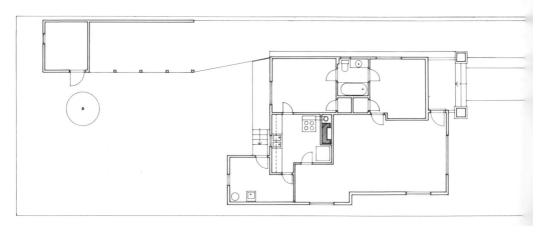

Figure 7.1 The Book of the House: view and plan of California bungalow, Berkeley, 1923. (Drawing by Lars Lerup.)

Figure 7.2 The Book of Architecture: Sebastiano Serlio, Theatre of Marcillis. (From *Sebastiano Serlio, Five Books of Architecture* [N.Y.: Dover, 1982]).

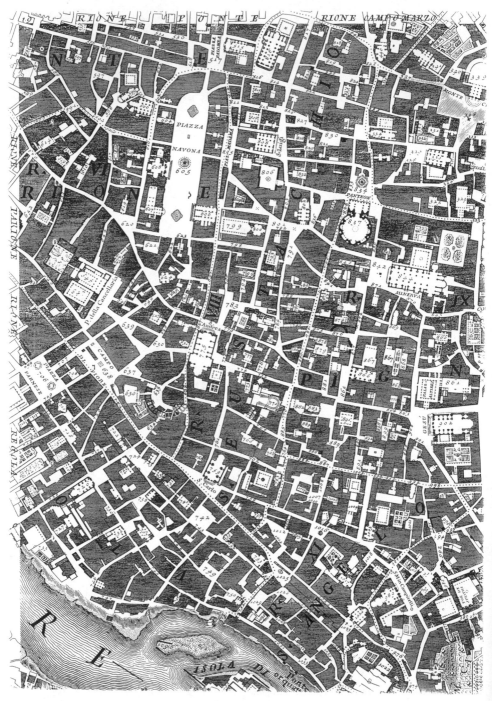

Figure 7.3 The Book of the City: Nolli's map of Rome, 1748. (*Nolli Map of Rome,* J. H. Aronson, Highmont, N.Y.)

Parallel to the idea of the house is an additional idea—the "book" of architecture. It is concerned with the fabric of the house, the very syntax of the form itself. For example, Tzonis and Lefaivre have defined three major concepts in their book on classical architecture: taxis, genera, and symmetry.[3] The history includes not only the house but palaces and temples, because the concern is not with use but with architectonic substance. Here resides the much contested autonomy of architecture.

An additional book casts its insistent shadow across the house—the city. The idea of the city is the most complex and obscure of the three, since we do not have a true theory of the city. Nevertheless, the city claims its place in the equation of the home, hinting at front and back, street, back yard, building type, and a way of life. Clearly, as an appendix to the city, suburbia warps the agenda, but the underlying idea of acknowledging a "community of houses" is shared by both.

The three books of House, Architecture, and City are too vast and complex to influence fully any one design; in fact, the best we can hope for is that all three can cast their shadow across the path of design. It is not true that the single-family house is the result of the idea of house only, but the influence of architecture and city are slight. The house design is hostage to an automatism that lays out the geography of each house with chilling predictability. An obsession with the fictitious dweller bars us from exploring the other aspects of the Idea of the House. The single-family house remains a caricature.

How can we circumvent this obsession to represent the dweller? How can we free ourselves from the insistent buzz of puppets playing house and allow the entire world of the house to enter?

Unhinging the Subject from Form: The Texas Zero

Texas Zero is a house that materialized in drawings, models, and texts. A fictitious client wanted a house of her own that would put a distance between herself and the single-family houses of her earlier life. The emphasis on this distance underlines the breach that always exists between dweller and house, one that is normally disguised by the mirror-play of the dweller and the inscribed subject.

Although the scope of this essay does not allow us to venture far beyond the Book of the House, Texas Zero deals not only with the complex shadow of the House but also with Architecture and City. The house sits on the frontier between country and suburb in Texas—between the past and the future. This has led to transformations of old Texas building types in an attempt to bring all three shadows to bear on the house.

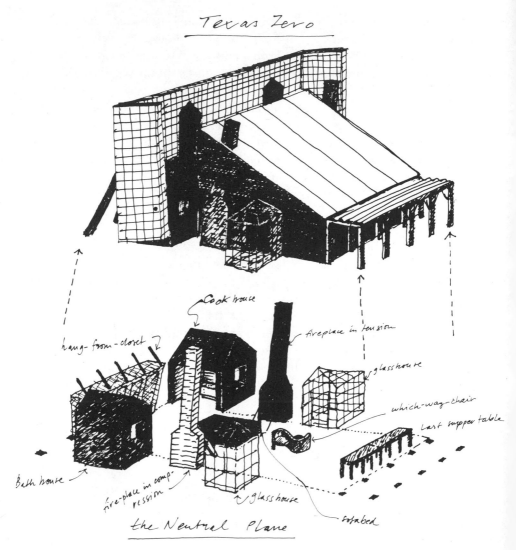

Figure 7.4 Texas Zero and the neutral plane. (Drawing by Lars Lerup.)

Texas Zero consists of several distinguishable elements. A large front wall (the country store) hiding two small black houses (the single-pen log house) is separated by a large loft space (the single-crib barn). Two gridded glass boxes, much like greenhouses, also flank the loft space. One end of the loft space abuts the back of the front wall and the other overlooks a valley, mediated by a porch (the front porch placed out back). There is no recognizable front door; only she (the client) knows where to enter. In fact, all openings in the five accessible spaces are enigmatic vis-à-vis their common names: door and window.

At the outset, the built object has distanced itself from the dweller.

The two mini-houses hold kitchen and bathroom equipment, respectively. Consequently, the typical kitchen occupies an entire house, small as it is, as does the bathroom. The whole house has been reduced to either bath or kitchen. Much like [black] shadows of their former selves, the two houses mark the end of an era—the tombs of a time when "houses were people"—and a return to when we knew better:

The Four Walls of My Room [4]

I know that all of them are poor
and that my friends deserved
other adornments, more distinctive
and more of them and larger ones.

But what do these words mean?
My walls have better manners;
and they do not love me for my gifts.
My walls are not like people.

Besides, they know that my things
will last only a moment
even for me. My joys and sorrows
and everything that I have done here

will pass quickly. The old
walls are indifferent to such gifts.
They are long-lived and they claim
nothing of my short life.

In the adjacent barnlike center space, a central axis that runs down the middle of the space is implied and establishes a datum for the entire project. The line is insinuated by the locations of elements such as the two symmetrically placed black houses and the fixed furniture. The central axis and these perpendicular axes of equipment and furniture form "new sentences," with implications for meaning. This domination of symmetry is the major shift away from the subject and its floor plan. The symmetrical placements work in insidious ways, both in and adjacent to the loft. The washbasin in the bathhouse is in the same location as the kitchen sink, as are the toilet and the stove, the bathtub and the dishwasher. These implied equivalences would be blasphemy in the single-family house, where bodily functions of one kind shall never meet those of another. Here "new sentences" of objects are formed.

Similarly, the fixed furniture forms new structures of meaning. A series of reversions is employed. A leaning fireplace in compression, a sofa that is also a bed, and another leaning fireplace in tension are lined up on a secondary axis that is perpendicular to the main one. As when the British satirist P. G. Wodehouse lets Bertie Wooster snarl: "He has, has he?" a reversal of meaning takes place. There, by using the symmetry of the comma, the statement of fact "He has" is put in question by the symmetrical reversion "has he?" The narrative flow is reversed and turned back on itself like a snake biting its own tail. In the Texas Zero the reversion has a similar effect, the very stability of the chimneys (underscored by their leaning) and the sofa/bed has been put in question. Yet their instability is only apparent and serves to reveal the source of their stability. The struts and tension wires reveal and figure the presence of gravity in the fireplaces, and the marriage of sofa and bed reveals their (un)comfortable similarity. Likewise, the [sleeping porch-which-way-chair-dovecote]-sentence suggests a new meaning (as does the last-supper-table as a miniature of the back porch). The nature of things is prominently displayed on their surfaces. Their form is displayed in contradistinction to their meaning.[5]

It is as if the checkered chess board has been erased and the pawns have been freed to invent their own arrangements. Like farm equipment in a barn, the familiar components of the house follow other laws than behavior. The new arrangement uncenters the subject, shifts it off base. A permanent parallax is established; the subject is unhinged.

The result is plan degree zero. The daily life of the subject, much like the narrative in a soap opera, is no longer represented in the syntax of the plan. The implied path from the front door to bedroom, with stops in the living room, kitchen, dining room, and bathroom, is erased. The freshly established order of familiar things poses alternative snippets of narratives that have both independent formal meaning and sociophysical meaning, the kind of tension that produces a stand-off, a status quo in which the dweller is at least momentarily freed to choose her own path.

Traces of a Sensual Subject: The Traditional Japanese House (and the Haiku)

If the subject in Texas Zero has been erased from the layout of the house, traces still remain in the equipment and furniture that are left behind. In the house described next, even these traces are at stake. In fact, the projection itself relies entirely on the "house" that the reader can "see" in texts and poems. The emphasis is on the Book of the House, but again the Book of the City enters in the complex dichotomy between house and garden— artifice and nature.

Roland Barthes writes hauntingly about Japanese space in his *Empire of Signs*:

> the Shikidai gallery: tapestried with openings, framed with emptiness and framing nothing, decorated no doubt, but so that the figuration (flowers, trees, birds, animals) is removed, sublimated, displaced far from the foreground of the view, there is in it place for furniture (a paradoxical word in French—*meuble*—since it generally designates a property anything but mobile, concerning which one does everything so that it will endure: with us, furniture has an immobilizing vocation, whereas in Japan the house, often deconstructed, is scarcely more than a furnishing—mobile—element); in the Shikidai gallery, as in the ideal Japanese House, stripped of furniture (or scantily furnished), there is no site which designates the slightest propriety in the strict sense of the word—ownership: neither seat nor bed nor table out of which the body might constitute itself as the subject (or master) of a space: the center is rejected (painful frustration for Western man, everywhere "furnished" with his armchair, his bed, proprietor of a domestic location). Uncentered, space is also reversible: you can turn the Shikidai gallery upside down and nothing would happen, except an inconsequential inversion of top and bottom, of right and left: the context is irretrievably dismissed: whether we pass by, cross it, or sit down on the floor (or the ceiling, if you reverse the image), there is nothing to grasp.[6]

This "house" stands free of any encumbrance; its slightly ghostly emptiness needs the presence of a subject that the haiku can provide.

The haiku and its tight economy of seventeen syllables allows the poet to capture both the Japanese world, the moment, and a limited set of "institutions." Parallels can be drawn between the house and the haiku: both are intentional constructions, "built" with economy and a purpose that combines a pragmatic concern for reality (in Suzuki's terms), artfulness or craftsmanship, and what the Greeks called *virtu* (roughly, virtuousness) or its Japanese counterpart, *satori*. And finally, more specifically connected with the intention in this text, the subject both in the haiku and in the house seems no longer to occupy the center.

> Completely imprisoned in the spring
> rain
> I am alone in the solitary hut,
> Unknown to humankind.[7]

The poet Saigyo huddles in his hut while the fine spring rain drizzles, but the hut is almost void because of his complete denial of his own ego. He seems to suggest that the insistent wall of rain cuts him off not only from others but from himself; what is left is the mere consciousness of the void and its wet surroundings. Perplexingly, and despite the strong presence of the narrator in the poem, the subject seems to be in the process of writing to erase itself. The reader is left with the smiters of words—a tear in his own everyday life—the rain and the almost empty hut. Surreptitiously, he has been placed inside the hut, behind the *amado* (rain door), which is slightly ajar. This door, or sliding partition between the inside and the world of rain, is only one of the membranes that he can put between himself and humankind/rain: the *fusuma*, the opaque sliding panel and its subunit the *byobu*, the folding screen; and the *shoji*, the translucent sliding panel covered with rice paper; and finally, the *nokisaki*, the zone of the shadow, cast by the overhang of the roof. The house has no windows in the Western sense, only layers that screen one place from another like filters. With a sliding motion the reader can open the amado, enter onto a veranda that mediates between hut and garden like a landing, and suddenly catch a glimpse of the "little frog," just about to jump into the old pond.

> The old pond, ah!
> A frog jumps in:
> The water's sounds! [8]

Barthes' gallery, Saigyo's hut, and Basho's pond begin to outline the parameters of our house—a house in which the body of the subject has been almost entirely erased; sensibility hovers in its place.

When Roland Barthes writes about the Shikidai gallery, "tapestried with openings, framed with emptiness and framing nothing," he hints at the house as frame, as scaffold, but also as a complex of tapestries, things in themselves, presences that bring the world to the dwellers' attention like the sound of a frog jumping into the "old pond." In the Japanese house there is an incessant partitioning, or framing that tugs at dwellers' gazes, frames their bodies, and marks their flesh.

The traditional house often has an area of stamped earth as part of the inner domain of the house; slightly raised above this surface is one made of wooden planks, with parts of it covered with one or more layers of mats. In time, probably by the twelfth century, the use of an entire floor of thicker grass mats called *tatami* (in 3- by 6-foot modules), came into practice. [9] Earth, wood, and tatami, all three corresponding to Japanese history from pit dwelling to sophisticated court life, exist side by side as an *aide-mémoire* of the history of Japanese dwelling. At least as important is the less European conceptualization of the sheer "suchness" of the earth: wood and grass under the hand, posterior and folded leg of the poet, shifting his weight from one to another while gazing through the framed view of Mt. Fuji beyond—Barthes' tapestries of thingness and alternating emptiness. The result is a house made up of frames ranging from the most ephemeral to the most robust.

Figure 7.5 Japanese space: Katsura Palace, Closet of the New Palace. (From Toko Shinoda, *Katsura: Tradition and Creation in Japanese Architecture* [New Haven: Yale University Press, 1961].)

Finally, because the house is almost complete, with its horizontal and vertical surfaces, the roof closes the interior of the sky, or attaches it, as in an old story of a couple who could not decide to mend the roof, because she liked to see the moonlight through the cracks and he liked to hear the patter of the rain. Saigyo was asked to resolve their problem as payment for room and board and he sang:

Is the moonlight to leak?
Are the showers to patter?
Our thoughts are divided,

And this humble hut—
To be thatched, or not to be
thatched? [10]

He was rejected. Without a direct connection, the Japanese house remains an eternal ambiguity (thatched or not, open or closed, inside or outside, house or hut) that leaves the subject adrift in its margins. The dweller is accommodated, surely, with much opportunity to tune the house like an instrument, but without figurative representation either of the physical or the social body (that remains firmly embedded in rituals only) and therefore left as another "intuition" next to frogs and stamped earth. As Barthes suggested, even when the house is viewed upside down, there is still not much that gives dwellers their bearings.

In the Japanese house the human trace has been transformed into architectonic substance, that is, the subject is not represented but present in the built itself. Unhinged, the subject's actions are transformed through construction into "shaped" material, such as the intricate wood joints, the *tatami*, the plastered wall, the *hinoki* cypress pole in the *tokonoma*, the sacred domain. As in the haiku, the substance of the house seems to lie in its reality and only initially in its image, which serves only as the *torii*, or gate, to a domain of intuitions.

> In Japan the problem of man is included, but in a philosophical way—he isn't approached on a physiological or social basis; instead a set of existential assumptions governs the subject. [11]

Conclusion

The ambiguous geography of the "neutral plane" and the sparse density of the Japanese hut and haiku suggest that the subject needs to be neither crudely geometricized nor tragically absent. However, to unhinge the subject as represented by "function" from form raises the specter of architectural autonomy, a subject that has been much misunderstood by many of the so-called humanistically oriented architects. Unfortunately, by now the lines between us and them are drawn rather sharply. Without repeating what I have said before about architecture's human endowment, the point here is rather simple and benign: most fundamentally, a building is nature transformed by labor through construction techniques. Architecture attaches itself to this fundament of building with the singular purpose of turning quantity into quality. Whether the two are ever separable is a question that must remain unanswered. What is clear, however, is that architecture is driven by a desire to improve the world. Since this utopian dimension is radically put into question by the threat of total annihilation, architecture may have to revise its own primary desire. In fact, it may have to announce solemnly its own death, although it may expect a long after-life.

Architecture's death is also its liberation from its ancient and arduous task to represent. For the first time architecture can afford to seek its own fortune. Its fortune lies just beyond the primitive hut, in the tapestry of the Japanese house, in the syntax of the Western house, in the gravitational field of architectural structure, and most certainly in directions outlined by many other architects. The common characteristic of all these directions is their marginal position vis-à-vis the ruling dogmas, and it is probably correct to suggest that to retain their vitality they must also retain their precarious status as an architecture of resistance.

Notes

1. Michael Foucault, *Discipline & Punish: The Birth of the Prison*. Transl. Alan Sheridan (New York: Pantheon Books, 1979), pp. 23–33.
2. The concept of the three books stems from a paper written about the central bank in Stockholm by the author under the title *I Staden, i Huset, i Banken . . . att lasa Celsing* (City, House, Bank) to read Celsing, Magasin Tessin, Stockholm, No. 1, 1981, pp. 17–35.
3. Alex Tzonis and Liane Lefaivre, *Classical Architecture: The Poetics of Order* (Cambridge: MIT Press, 1987), pp. 1–6.
4. C. P. Cavafy, *The Complete Poems of Cavafy*. Transl. Rae Dalven (New York: Harcourt Brace Jovanovich, 1961), p. 228.
5. For a more complete description see the author's *Planned Assaults* (Montreal and Cambridge: Canadian Center for Architecture and MIT Press, 1987), pp. 81–92.
6. Roland Barthes, *The Empire of Signs* (New York: Hill and Wang, 1982), pp. 108–110.
7. T. Daisetz, *Suzuki, Zen and Japanese Culture* (Princeton, N.J.: Princeton University Press, Bolingen Series LXIV, 1959), p. 341.
8. Ibid., p. 227.
9. Marc Treib, *Domestic Implications*: The Traditional Japanese House (1982, unpublished), p. 7.
10. Daisetz, *Suzuki, Zen and Japanese Culture*, p. 340.
11. Chris Fawcett, "An Anarchist's Guide to Modern Architecture," *Architectural Design* (1978):37–57.

THE SUBJECT'S IDENTITY

Lerup leaves us with questions. Who are these people inscribed on structure and how did they get there? For whom do they exist? Johanna Drucker answers, finding an adaptation of the concept *subject* from literary criticism in the works of Lerup and Peter Eisenman. She traces the workings of the subjects in their radical designs for the Nofamily House, and House X respectively. Critical inquiry into the subject wrests us from simplistic notions of user, client, and architect. To quote Drucker, the subject is "produced by the architecture itself in a way which parallels the production of a subject of a literary text."

Lichez offers an abundant imaginal resource directly from literature, particularly fiction. He suggests that since writers' people are so carefully wrought, we can teach young architects to create the subjects' identity through literary models. Presenting quotations as images in their own right, he shows us the writer's techniques for designing habitable places. We are reminded of Joe Esherick and Robert Kliment, who, among others, have found inspiration in novels for shaping their notions of people in place.

Examining a very different literature, Juhasz explores myth, the social sciences, and psychoanalysis to discover conceptions of human nature and identity in Western thought, and how these have converged with varying conceptions of architecture. Theories of human nature (existentialist, behaviorist, biological) give way to mythic distinctions of theme, structure, and identity that frame the discovery of architects' people.

These chapters explicitly search for a deeper understanding of the hidden actor's identity and the role they might play in design thinking. Recalling Schutz' critique of the sociological homunculus, the present authors step outside of architecture to find the appropriate frame, or living locus, for designers' mental dwellers. To structure these variegated images, each of the following chapters redefines or expands the notion of architects' people, allowing us to analyze architectural works and ideologies in a new light. Each author in this section uses some form of text as a heuristic resource pool. The texts include language theory, myth, and literature—each offering a different path to the sources of identity. With these texts and the larger perspective they offer, the authors examine specific architectural activity.

8

Architecture and the
Concept of the Subject

JOHANNA DRUCKER

It has become increasingly common over the last few years for concepts and vocabulary generated in one discipline to be borrowed and applied to another. The attempt to apply the concept of the "subject" to the practice of architecture is an example of such borrowing. Originating in fields beyond the normal scope of architectural practice, such as literary criticism and psychoanalysis, the concept of the subject stands opposed to that of the person. Where the person is generally assumed to be an individual human being whose existence is independent of any particular architectural form, the subject is considered to be produced by the architecture itself in a way that parallels the production of a subject of a literary test. The evocation of the notion of the subject by practitioners Eisenman and Lerup suggests that it may have useful implications for the conceptualization of architectural projects as well as for their execution. This chapter will discuss the ways Lerup and Eisenman have appropriated the concept of the subject, and the distinction between subject and person will be examined for its relevance to an architectural practice.

The concept of the subject has evolved in linguistic, psychoanalytic, and psychoanalytically influenced literary criticism for at least the last twenty years. Peter Eisenman's project House X and Lars Lerup's Nofamily House both use the concept and examine its appropriateness to architecture. They do not represent two typical cases out of many; these two examples stand almost alone in their application of the concept to architecture. Fortunately, they also stand in a relation to each other that permits contrast and comparison. Each uses and defines the architectural subject differently; each appropriates from similar but slightly different disciplines and sources. Separately, each offers a case study in the use of the subject in architectural practice; together their work begins to define a territory to be investigated by further exploration.

Before launching into the examination of Eisenman's House X and Lerup's Nofamily House, it seems useful to sketch the origins of the concept of the subject within the disciplines in which it has been developed and to differentiate among the various formulations of the concept, noting the distinctions that will also differentiate Lerup and Eisenman's uses of the notion. Second, it seems worthwhile to situate Lerup and Eisenman within the development of critical theory in general (again, largely with respect to its literary, linguistic, and psychoanalytic origins).

The subject has its origins in the theoretical domains that concern themselves with the operations and functions of language. In the course of its recent shift from concern with the text as an autonomous entity, literary criticism began to recognize the necessity for considering the ways in which a text constructed a reader. This reader, sometimes considered an *ideal reader*, was not so much a person responding to the rhetorical strategies and thematic content of the text as an actual *product* of the text. A distinction between the *person* who might read a text and the *subject* produced by it was established. This distinction will be exactly the one I will use here to distinguish the concept of a subject of architecture from the conventional idea of a person.

A person, the reader of the text or user of a building, has a social profile, a personal history, physical needs, and an idiosyncratic (if often typical) attitude toward the architectural structure. The person is separate from the work, has an autonomous existence, and comes to the building with his or her own history, prejudices, and expectations. The subject, on the other hand, has a structural origin within the architectural form itself. In textual analysis, the subject arises from the organization of the literary text: from its voice, point of view, narrative strategies, in fact, all the features of the text that posit a relation among its elements and use those relations to situate the reader in relation to itself. The concept of the subject is intimately connected to the structural features of the text. In fact, the subject is made, given form, through the projection or ensemble of those features. In its simplest, most basic form, the subject should be understood as the product of the text and as totally dependent upon the text for its existence. (We will elaborate on this concept later to demonstrate some of its complexity, especially as this is relevant to Lerup's project.) The person, on the other hand, is an autonomous being, an actual reader.

It should be clear from the preceding that the person and the subject are completely different from each other, almost opposed to each other with respect to their origin and disposition toward the work—building or text—

to which they relate. Unfortunately, this distinction cannot be as nicely maintained in practice as it can be formulated in theory. The interaction of the actual reader with the text, or of the user with the building, will involve some negotiation of the individual person with the subject that is projected from that text or built form. It is precisely the ways in which Eisenman and Lerup differ on their evaluation of this process of negotiation that permit a distinction between their attitudes toward the subject to be examined. Both have different definitions of the subject. Both have different means of arriving at the way they believe the subject to be formulated in the architectural practice. An elaboration of concepts of the subject requires sketching the background of critical theory against which the definitions operative within Eisenman and Lerup's projects are figured.

The initial concept of the subject as a product of the structural features of the text found support in the linguistic analyses of critics such as Emile Benveniste and Gerard Genette. Concerned with the specific features of language that determined the relation between a text and its reader, these critics analyzed the use of pronouns, voice, forms of address—in short, any of the features that organized the point of view *of* the text with respect *to* the reader. This critical position, incorporated into literary criticism, was consonant with the aims and endeavors of what is termed a structuralist position. Although it elaborates slightly on the earlier, more rigidly defined structuralism of a first generation including Claude Levi-Strauss, Ferdinand de Saussure, and Roman Jakobson, its methods of analysis are closely linked to the investigations of these writers. The key emphasis of the structuralist position was on the way in which a particular system (for example, language, kinship, and poetics) functioned. The elements within the system (such as phonemes, relations, and rhyme) gained their value largely in relation to each other, rather than from any independent, essential qualities or intrinsic values, such as sound or blood links or transcendent symbolic meaning. Within this particular analytic mode the attention to formal features of whatever was under investigation became the primary emphasis of the critical technique. Although the leap from the study of language or culture to the study of architecture is a large one, it is easy enough to imagine the appeal of such a structural approach to the examination of architectural practices. This approach is very close to that used by Eisenman, not just as a critical practice, but as a generative one. Eisenman's disposition, for the sake of this discussion, can be characterized as structural, and his concept of the subject was largely formulated within this classic structural mode and attitude.

The methods of structuralism, as the term implies, have as their limit the structure of the system being examined. The structure is assumed as a kind of empirical fact, or at least, cultural and social fact, and the individual is assumed to operate within that system without too much influence on the choices imposed by its order. The limitations of this approach for the study of texts quickly became apparent when the examination of language within the psychoanalytic framework began to be brought to bear upon the study of literary language. A Freudian psychoanalytic conception of the work of language, with its clear division between latent and manifest content, its operations of displacement and condensation, its work to recuperate the repressed into an articulated text all figure into a fairly orthodox structural mode. But the reworking of this Freudian concept within the work of the radical French analyst Jacques Lacan helped to formulate what came to be known as a poststructural position. Within this position the subject has a more complex relation to language and the emphasis moves away from structure per se to *structuring* as a *process*. It is this conception of the subject that Lerup uses in his Nofamily House.

Poststructuralism is largely concerned with examining the ways in which disciplines, systems, and discourses are manifest and articulated. Rather than assume the system, as in a structural mode, poststructuralism examines the assumptions and processes by which such a system can be conceived. It aims at deconstruction, taking apart and examining processes of support and structuring, that make possible most of the institutional orders within which we conceive of and organize our lives. Foremost among these is the order of language. For the poststructuralist and for Lacan, Derrida, Foucault (though they might deny such an affiliation, their methods are a demonstration of poststructuralist approaches), showing the ways in which systems of order conceal their mechanisms of control or effectiveness is fundamental to understanding their cultural function. For Lacan in particular, the study of language becomes a study of the ways in which the acquisition of language inscribes the individual within the social order both by the thematic organization of language as a system and by the process of its acquisition. Lerup will attempt, as I will show, to make a parallel analysis of architecture, emphasizing the two aspects: the content of architectural "vocabulary" and the structuring process that architecture effects upon an individual experiencing built form. Whether the parallels between language and architecture can be sustained is a matter of some question; the attempt to sustain them provokes at least an interesting discussion of the implications of architecture as a discourse and discipline.

Since this Lacanian notion of language is fundamental to Lerup's work, it needs some elaboration. For Lacan the subject is not simply an entity brought into being through the structure of language, not simply a projection of a text and of its specific rhetorical and linguistic strategies. The subject has an evolution that begins in infancy and the subject is contin-ually formulated in relation to language in an ongoing process. The evo-lution of the subject goes through several phases. Initially, according to Lacan, an infant exists in a prelinguistic state without a clear sense of its own unity as a being. Its body is confused with the mother's body and it has no idea of separateness as an experience or a concept. With the acqui-sition of language, in what Lacan terms the "mirror stage," the infant be-gins to enter into the symbolic order of language simultaneous with recog-nizing the autonomous unity of its body through its image in the mirror. Language is considered overwhelmingly patriarchal, not only in its the-matic content, but also in its imposition of dominant hegemony as an order and law. These themes will also be taken up by Lerup in his analysis of architecture.

Finally, for Lacan, the subject's relation to language remains one char-acterized by feints and fictions, deceits and simulacra, all the pitfalls and problems of representation. For the subject, once initiated into language, constantly articulates itself, and this articulation, to state this in typically Lacanian doubletalk, is always exactly what the subject is not. Very simply, language—the representation or articulation of the subject—is a foil, a guise created by the subject, who continually escapes the very definition offered by that articulation. On the other hand, the subject has no existence for Lacan outside of language, and therefore the process of analyzing the sub-ject is one of continually analyzing the subject's language in search of the elusive presence of the subject itself. When Lerup describes architecture as a symbolic system, he is comparing it to language as a symbolic system in the sense conceived by Lacan, and Lerup's subject goes through many of the same kinds of gymnastics and sleight-of-hand maneuvers as Lacan's. Although architecture cannot be demonstrated to function or be formed like a language, it can, within Lerup's example, be shown to partake of some of the features of an articulate discourse.

Although this may seem like an inordinately lengthy preliminary to the discussion of Eisenman and Lerup's work, situating the concept of the sub-ject within these structuralist and poststructuralist critical frames should provide the necessary context within which the concept can be usefully compared to that of the person and its application to architecture examined.

Peter Eisenman's approach to the process of design in the project House X was clearly informed by an acquaintance with structuralist methods and criticism. It represents a continuation of earlier work in which he attempted to demonstrate the possibility of applying linguistic theory to the practice of architectural design. As described by Mario Gandelsonas, in his introduc-tion to *House X*, this earlier work attempted to produce a systematic organization of the codes of architectural practice, to define an apparently finite and stable number of forms and their correlated meaning within a closed system, that is, to catch the illusion of a language.[1]

The attitude toward language suggested by this approach is classically structuralist, based entirely on Saussure's definition of language as a closed system of finite elements mutually determining each other's value. The appeal of this definition of language as a formal system (the limitations of which have long been pointed out and superseded within the study of linguistics itself) is that it seems to offer a ready analogy for architecture. Eisenman's earlier projects (for example, House I and House II) had been demonstrations of the effect of borrowing this concept of language as a model for architectural practice. A finite set of elements was manipulated through a fixed set of operations in order to generate spaces and forms through a logical process. Eisenman's aim in doing this was to avoid the inherited modernist formula of determining form from function. Form was to be generated by formal means, and the "language" of architecture was to refer to itself rather than being determined by external concerns or needs.

Eisenman conducts his project within the language of drawing. He sets up his initial vocabulary by establishing a form, initially a square form, and then transforming it in a sequence of stages. He divides it, changes the orientation of the quadrants, shifts their alignment along the lines (literally drawn lines) of their division, allowing breaks generated in lines by the incidental fact of their length to become openings into the walls represented. An element such as a small unit of drawn grid makes its appearance, is repeated, and comes to stand for a form. An orthographic projection evolves from the original plan, itself formed from this process of repetition, transformation, always conceived without regard for the functional aspects of the forms being generated and with careful attention to their aesthetic relation as forms. Function is subordinated to form in this process, not allowed to assert its hegemony over the design as it evolves in the manipulation of formal elements. For Eisenman the elements are the "vocabulary" and the transforming operations are the "syntax" of the architectural language.

There are obvious problems with the analogy between language and architecture, especially made on this kind of formal basis. The symbols of language are entirely arbitrary, though they enter into constraints once they are operative as a system. The elements of architecture are not, in a strict sense, symbolic, but have structural and operational constraints from the outset that restrict, if they do not entirely determine, their form and their relations. These early projects need to be seen as exercises, however, in the generation of architectural structures from a formal approach to architectural elements, and not faulted on the obvious impossibility of maintaining the language–architecture analogy.

Eisenman takes his investigation of language one step further than the rigid Saussurean language-as-closed-system approach and delves into the generative grammar of Noam Chomsky. Chomsky's work is frequently perceived as an extension of a certain line of structuralist thought with respect to language, and the extension of Eisenman's thinking along these lines follows consistently with his beginning approach. The introduction of ideas from Chomsky's system allows Eisenman to propose rules for the generation of architectural systems similar to those being proposed by Chomsky as basic to the generation of linguistic statements.

Chomsky's generative grammar was based on the idea that there was a distinction to be made between what he termed the deep structure of language and the surface structure. The deep structure contained kernels of language, semantic and syntactic elements in raw, unrefined form. These were transformed through a series of operations and refinements into intelligible statements. The rules that governed these transformations were the object of Chomsky's research in the generative grammar project, which ultimately failed. The appeal of the idea, again, lies in its formality, in its suggestion of a system that might be fully self-contained and self-defined, governed by an identifiable and finite set of rules operating with precision and control over a finite set of elements. Adopting the look and vocabulary of Chomsky's system, Eisenman identifies deep structure elements in his architectural system; designates the processes through which he alters, multiplies, and refines them; and considers the final product a surface structure architectural statement.

House X essentially continues this process without fundamental reassessment of either the linguistic models being used or the fallacies or at least difficulties of the analogy on which Eisenman had premised his project. But a few crucial elements in House X distinguish it from previous projects, not the least of which is attention to the concept of a subject in linguistic terms. Eisenman also seemed to have exhausted the possibilities of working within the limitations of a purely logical system of operations and procedures.

It is important to remember that until this point in his work, Eisenman evidenced little consideration for the user of his architectural forms. A significant aspect of this consideration is that we have largely been concerned, in the work of Eisenman, with the examination of drawings rather than built forms, or, more specifically, with the generation of drawings rather than with the experience of a built environment. It is true that in the final stages of his design process, once the full complement of spaces and struc-

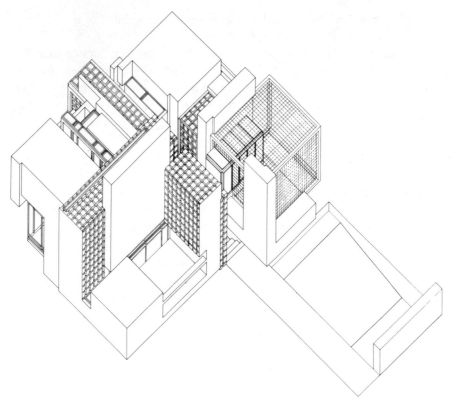

Figure 8.1 House X by Peter Eisenman; axonometric. (Eisenman Robertson Architects.)

tures is in place, fully generated by the transformations of original elements into the surface structure of final form, Eisenman has the uncanny habit of allowing the spaces to be tagged with names designating their function. Suddenly the words *bedroom*, *kitchen*, and so forth, appear like so many revived corpses waiting in the wings for the moment of their resuscitation. But the designation of use value is still far from the investigation of the person or user as an element of architectural design.

As far as Gandelsonas is concerned, however, it was necessary for Eisenman to reach this point of realizing both the full extent and the resulting limitations of his project before he could enter into consideration of the subject of architecture.

> At the point when this *object* (architecture) becomes clearly and almost autono-
> mously defined in its systematic, internal, formal relations then does the *subject*
> take on a clear configuration. In linguistic terms the definition of an *organization*
> as a normative system, which in architecture would be the constitutive rules of
> the *object*, implies at the same time its *subject*.[2]

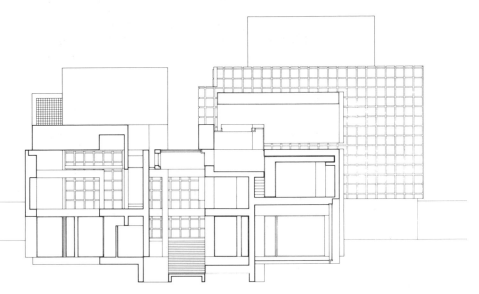

Figure 8.2 House X by Peter Eisenman; section. (Eisenman Robertson Architects.)

The subject implied by Gandelsonas' assessment of Eisenman's work also bears little resemblance to a person or user. The subject proposed as the extension of an autonomous, rule-bound system is highly idealized. This is the subject of discourse, a subject implied as the necessary user of a system of rules or of a system operated according to procedures that remain unused without the presence of a subject capable of making them function. In a Saussurean conception of language, this subject is implied as a speaker, the entity whose use of language allows for a distinction between *langue* as the system of formal elements and procedures and *parole* as the individual use of that system for actual speech. The distinction between these two, a part of Saussure's original formulation, implies a subject, a user. It is this rather limited concept of the subject that Gandelsonas discerns in the work of Eisenman. The effect of individual character, of life situation, of personal needs, or of any of the rest of the features generally associated with an architectural client or person remains conspicuously absent from this notion. And not without reason. For Eisenman the house, House X, still finds its identity as a result of operations carried out within the formal vocabulary of architectural form as form.

The insistence of Gandelsonas on the notion of a subject in the House X project can be explained by the feature that distinguishes House X from the other Eisenman House projects: namely, that element of the irrational included in the procedure. With the preceding projects, Eisenman, in his attempt to discover a consistent language of rules, elements, and transformations that he could clearly name an architectural language, had been faithful to a process he termed *logical*. In this process his operations had been repeatable, his elements had been duplicated, his transformations had been reversible and recoverable. The final form of the Houses, the so-called surface structure form of the architecture, had retained the traces of the process by which it had been achieved. Its form had a certain discernible degree of history in it. The rules could almost be discovered in the changed relations of the elements from their original positions to the positions in which they were found in the final stage of the project. With House X this apparent recoverability of rules, elements, and transformations is no longer a feature of the final product. This is the result of the fact that at every stage of transformation Eisenman includes some idiosyncratic alteration of the process, element, or rule. There is a random, or at least inconsistent, aspect to the process that confuses the history and makes the result a more complex combination of consistent and unpredictable methods.

It is only fair to assume that Eisenman's decision to introduce these random elements was an assertion of his own idiosyncrasy. It is the recognition, or at least the expression, of individual personality, almost an act of rebellion against the rigid formality of the processes he had employed until that point in the House projects. Having attempted to discover a coherent, systematic architectural *langue*, he was finally prepared to risk testing the parameters of an architectural *parole* by manifesting his identity as the user of the system. It is in the name of individuality that the assertion of his own presence as a subject within the architectural discourse he is constructing makes itself felt. This individuality, which assures itself of identity by recording its own idiosyncratic character as a detectable trace in the architecture, conceives of itself as autonomous, intact, already in existence before entering into its use of language or architecture. This is a somewhat romantically conceived individuality and far from the subject that will be discussed with respect to psychoanalytic theory. The important distinction, and one that distinguishes Eisenman from Lerup, is that this individual identity does not depend upon its interaction with a symbolic system in order to be produced.

The subject in Eisenman's architectural language is a long way from a user or person, but the reinvention of its presence in House X at least signaled the return of imagination to the rules of an architecture conceived to demonstrate its possible consistency as a language. If that original project seems difficult to justify, at least in the sense that it is so artificially conceived, it should be remembered that the essence of the project was to return to architecture its own specificity, to understand it from within its own practice as a systematic and structured discipline. If Eisenman's project ultimately demonstrates the full extent of the difference between architecture and language, it nonetheless confirms the actual character of architecture within the negative construction of *how* and *why* it is *not* a language or even *like* a language. And it also confirms the subject within the discourse of architectural practice, in this case, the evident and essential subject of the architect. The work of Lars Lerup, on the other hand, moves into very different formulations of the language–architecture analogy and raises the specter of an entirely different kind of subject.

In reviewing Eisenman's project House X, Lars Lerup makes a somewhat different assessment of the subject, one that Lerup proposes as a projection of the physical form of the architecture:

> It is important to note that the critic or reader constructs this subject on the basis of projections from the house or text. It can be likened to a hologram whose source is the representation of the house (or the house itself), as opposed to the architect or the actual human subject (the user). The placement of windows, doors, and stairs, the layout of rooms, orientation—those all imply a subject that may not coincide with the actual user, and which the critic may not see as the architect intended it to be seen.[3]

Lerup's notion of the hologram subject as a production of the house form can be compared to the idea of the subject produced by a literary text. This is the second definition of the subject that we have encountered. A third, psychoanalytic subject will also emerge with respect to Lerup's work. In literary theory the notion of the subject as a product of the text became a useful concept insofar as it allowed for examination of a number of levels of organization of the text itself. It provided a means of talking about the separation between the point of view of the narrator and author as located in the organization of the narrative, of identifying the way in which the text positioned the reader with respect to various rhetorical and structural devices or points of view, and of articulating in sum the various strategies through which the text posed a relation with the reader. The theory evolved from the application of structural, linguistic analysis to literary texts, and its application to other realms of discourse is less simple to construct. Once again we are faced with the almost insurmountable problem of the differences between discourses—in this case between literature and architecture. The linguistic nature of literary discourse allows for the ready identification of elements like voice, point of view, tense, pronouns, and so on, all of which organize a distinctly articulated subject and subject relations. The nonlinguistic and in many ways nonsystematic (or at least, inconsistently symbolic) quality of architecture makes the search for this subject more chimerical. Where is the point-of-view system of a building to be located?

How is the use of voice to be identified? And if these linguistic elements are not duplicated within architectural discourse, how is the architecture to be "read" with any predictable consistency?

There are no evident solutions to these problems. Lerup nonetheless is interested in pursuing the approach as if its legitimacy can be assumed. The suggestive possibilities that emerge from this approach are numerous, and even if they emerge from murky territory, possibly unsupportable under scrutiny, Lerup uses them to interesting ends in his own work. It is difficult to find any support for the assessment of Eisenman's work in these terms, and I would suggest that Lerup is here already justifying his own experiments in projecting this assessment onto Eisenman, rather than truly extracting this reading from House X. At least, Eisenman gives very little indication that he is considering the ways in which the generation of form through logical processes necessarily results in the production of such a holographic subject. He seems relatively unconcerned with postulating such a subject relation to the discourse of his architectural project and in no place articulates relations among elements such that they demonstrate a self-conscious discourse, one aware of its production of a subject. Although Eisenman's work is available to such a reading, as, in Lerup's assessment, *any* architectural work would and must be, Eisenman's own discussion of the project does not seem grounded in the consideration of this particular subject effect.

However, this hologram subject plays a substantial and conspicuous part in Lerup's Nofamily House.[4] In this project Lerup also explores the possibilities of posing the same relation between architecture as a discourse and a subject as that which is considered to be generated between language and a subject in a psychoanalytic formulation.

Lerup's project has in common with Eisenman's a conceptual base that functions as a kind of operational premise. By making certain theoretically based decisions at the outset, Lerup works through the design process according to procedures that have both a conceptual justification and an architectural effect. As did Eisenman, Lerup works largely on the drawings and within the vocabulary of drawing. Many of his conceptual notions depend upon drawing in order to be carried out—the notions of erasure, for instance, or even duplication or change in orientation—and though they result in physical, architectural effects, they are carried out within the parameters of the drawing medium. But where Eisenman was concerned with the establishment of (or discovery of) a set of rules specific to architecture that would allow it to function as a language, Lerup is more free in his borrowings from a variety of sources and disciplines. He forces architecture to submit to a number of theoretical operations and reconsiderations, some of which are linguistically based, some of which are merely most readily described in terms of language, and some of which bear little relation to language at all, in fact, work to emphasize the autonomy of architecture from language. Because of this, it will never be simple to identify the hologram subject of the architectural practice that Lerup so gracefully described in discussing Eisenman, since Lerup's architecture is continually subjected to processes of construction and deconstruction, of building and breakdown, so that any subject whose existence depends upon form would be frequently at the mercy of such transformations. Nonetheless, this subject serves a constant, persistent purpose for Lerup, which is to ground his architectural practice in a self-referential frame. Rather than extend himself to an imagined user, a necessarily fictive individual, Lerup continually returns to the architectural form and questions the subject it might produce at any particular moment of construction. This subject provides Lerup with the means of examining the effects and structures of his own practice.

The Nofamily House project involves proposing a construction on a site with already existing structures. These structures were perfect archetypes of the family house with wood frame, peaked roof, front stairs, and so on. One of these was too badly deteriorated to be left standing, and Lerup emphasizes this dramatically by erasing all the matter from the drawing of this structure. Devoid of material it stands as a figure, and that figure can be read as the inscribed discourse of the family. The entire project revolves around a deconstruction of the family and a subversion of its inscription upon the structure of the house, and of the family as it is assumed to underlie the typical house structure.

For Lerup, everything the family is or does as a network of interactive functions is compiled as the "trace of the family," which is the figure of the house. Throughout the project, Lerup uses a verbal narrative to complement and form a dialogue with the drawings, and in the process he establishes linguistic points of reference as well as graphic ones, allowing the two domains to interact and to point to and away from each other. He identifies "figures" in language and then points to their inscription in the drawing and, similarly, allows images that surface in the design process to become identified with linguistic terms. In so doing, Lerup risks confusing the two domains, but he also allows for a certain strategic discussion in which architecture takes on linguistic characteristics, not, as Eisenman sought, by constructing a strict analogy based on a structural model, but by making momentary equivalences or definitions. Thus, concept such as the "trace" of the family or the "figure" of the house come to operate in both the drawings and discussion. Similarly, Lerup uses the term *reading* loosely and figuratively in assuming the production of a subject by an architectural "text" without taking the process too literally.

Lerup is certainly aware that the verbal characterization of the house encodes a pattern of thinking that is not separate from the architectural practice in which the house exists. Clearly, the subject produced by architectural discourse is also a subject informed by the cultural practices of language. As he proceeds in the deconstruction and reconstruction of the original house, transforming it from a typical family house into the new form of the Nofamily House, Lerup attempts to construct a situation that confronts the reader with the assumptions brought to the house in the conventions of this normal language. He does this by making spaces that cannot be, as Eisenman's were, immediately renamed with conventional terminology of *kitchen* or *master bedroom,* and by obstructing the conventional "narrative" of the house itself through systematic attempts at subversion. These will be best understood, as will their role in relation to the production of a reading subject and a psychoanalytic subject, within the overall description of the process of the house's full design.

From the point at which Lerup erases the substance of the first house, letting it stand as a figure against the body and substance of the second, he initiates a process of deconstruction that attends to the very form it is undoing. Lerup's is a calculated subversion, attentive to the architectural process as well as to the reading subject that will perceive and be produced by the designed form. Almost immediately, Lerup introduces elements of a psychoanalytic discourse into the series of operations he performs on the deconstructed house. Initially, he erects a glass house that by its transparency, allows the figurative form to be perceived without giving it substance. Lerup is conscious of puns and parallels in his choice of words, attentive to their architectural reference and their significance in the jargoned vocabulary of critical discourse. Thus, the use of the glass is a comment on the "transparencies" of language as a medium that has the appearance of being merely a means of describing figures, although it is actually the sole means of formulating them.

In the next step, Lerup doubles the glass house, duplicating in the development of the structure a phase identified in the psychoanalytic theory of Jacques Lacan as the mirror phase. Lerup wants us to believe that the house confronts itself, forms itself, discovers its identity, in relation to its mirror image in the same way that an infant in the mirror phase constitutes itself with respect to its reflection. On the one hand, this is a reductive (possibly even to the point of being incorrect) reading of Lacan's rather more complex argument. On the other, it is Lerup's attempt to formulate

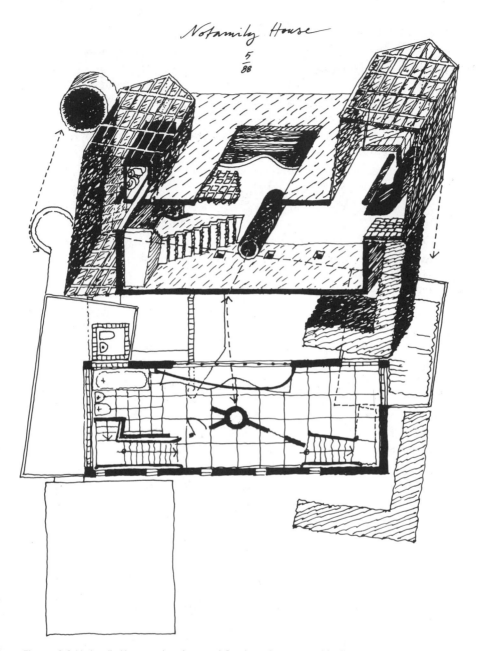

Nofamily House
5/88

Figure 8.3 Nofamily House: plan (second floor) and axonometric (from under) combined. (Drawing by Lars Lerup.)

a figurative relation between design as a practice and architecture as a symbolic discourse within which some of the same functions might be performed as occur in language. Arguably, Lerup's presentation is a demonstration of the verity of this proposition. Obviously, architectural design does function for Lerup as a symbolic discourse, one in which the same kinds of problematic confrontations, formulations, feints and dodges, fictions and discursive practices may occur as occur for him in the more readily recognized symbolic order of language.

Lerup thus deliberately confuses architecture and language, first by allowing the name of architectural elements to be used in a linguistic discussion as if they were the thing and to allow properties characteristic of the linguistic term (e.g., *figure*) to return to the architecture as if there were no problem in interchanging the two. Basic issues, such as whether a particular configuration of walls, floor, windows, roof, and door constitutes a figure, are glossed over for the sake of the use that can be made of this readily available analogy. More important, Lerup wants to allow architecture to serve as a symbolic order capable of encoding and inscribing a subject, a psychoanalytic subject whose only identity is that which the symbolic order produces. Again, the possibility for architecture to sustain a *reading* or *articulation* within its material form in the same way that language does is questionable, but posing this possibility permits architecture to take on the role of an active participant in the psychoanalytic process. Rather than being considered the image or expression of a particular psychological position or cultural form, the architecture functions as a means of formulating such an image in a dynamic relation with an active subject. The architecture–subject relation is a vital one in which both aspects are essential to the formation of expression and language conceived as a symbolic system that does not exist independent of its relations with a subject.

Lerup is attempting to layer several kinds of subject relations into the Nofamily House project. Initially, by dealing with architecture and language as an interrelated system he points out the extent to which the two encode similarly constructed cultural subjects whose expectations, desires, and needs are far more produced and far less individual than many would like to believe. In this sense Lerup participates in a general critical tendency that questions and negates the traditional humanistic concept of the person and seeks to replace it with these various categories of the subject. Finally, he also attempts to develop the house through a set of procedures modeled on the stages identified in psychoanalytic literature as those that lead to the formation of a psychoanalytic subject through the acquisition of language. For the psychoanalyst the subject exists only in relation to language as a symbolic system in which he or she is articulated as a subject. Positing that architecture can also function as such a reflective system, Lerup attempts to let architecture both function as a symbolic system and mimic the specific characteristics of the symbolic order of language as it is conceived in psychoanalysis.

Consequently, Lerup allows the Nofamily House design to transform itself through operations that become associated with terms like *auto-affection* or *hetero-affection*, which Lerup uses as if they are forces capable of transforming the relations between architectural elements. Again, by allowing the drawings and the language to become involved with each other, he can operate on a figurative level that slips between the one realm and the other. Similarly, he identifies the central feature of the house, a column, as the "anthropomorphic trace," and by so naming it, he allows a number of significant points to accrue to its presence. For instance, the whole figurative discussion of architecture as a symbolic system capable of producing a reading subject, a holographic subject, or a psychoanalytic subject depends upon the belief that a reader, viewer, or designer could *identify* with architecture in the same sense that she or he could identify with a construction in language. The central position of the anthropomorphic trace of the column in the Nofamily House is first of all a comment upon this necessity for a point with which to structure a primary identification. Second, by its position, the column prevents the subject (whichever subject) from being able to occupy the very place on which its identification is based. That is, it cannot become the very thing with which it identifies. This is necessary, since the subject cannot be its symbolic construction and cannot be made equivalent to either the identity constructed in language or that constructed in an architectural discourse, since both function as representations to the subject, structuring the subject within their own discursive practices.

The central column has another aspect to it: It functions as a phallic object at the center of the house concretizing *phallocentrism* into architectural form. The image and the terminology it requires connect Lerup's involvement in linguistic issues to the metacritical level of conceptualization. At this level language is examined for its own biases, prejudices, and underlying philosophical assumptions, of which the notion of an equivalence or relation between logocentrism and phallocentrism is one of those called into question by contemporary critical discussion. As a phallocentric object the column functions as a sign of patriarchal power, an immovable point of reference in the site, a final referent of patriarchal authority inscribed within the discourse of the family. Although this formulation and the particular political discourse in which it participates are open to question, Lerup's use of it here simply demonstrates the extent to which Lerup's psychoanalytic propositions are idiosyncratic, individual ones in which his own conflicts and issues are manifest. The Nofamily House, like the Love House and other of Lerup's projects, epitomizes a particular psychological issue in which Lerup is bound up. He demonstrates by this body of work the actual extent to which the practice of architecture necessarily works out his own subjective issues in the relation he forges with its symbolic order.

Here we return to Lerup's fundamental recognition that language and architecture are not the same while working to demonstrate the complexity of their interrelation. For instance, there are several "traps" in the house that serve as the means of making one aware of its form, its materiality, and of forcing the question of that form at a moment when there is no familiar or adequate language to negotiate the encounter. The forms in these traps do not operate according to their usual, named function. For instance, a stair railing, detached from the stairway, becomes a thing in itself, no longer reassuringly defined as functional. Its arbitrariness is revealed so that it is unrecognizable except in itself, by itself. The materiality of the thing is made apparent. Although Lerup is in no way systematic, he is close to Eisenman's project of attempting to demonstrate an architecture as such, one remote from definitions of functionality and grounded in form. The difference is that Lerup does not strip the architectural elements of their figurative value; rather, he allows those figurative aspects to be perceived outside of their usual functional definition. He is well aware that any form will invoke history and what he terms the *hermeneutic phantom*, full of referential images, specific tales, myths, and characteristics, which hangs onto any bit of material that takes shape. There is a strong connection between this historicity and the language in which the forms have been defined and redefined through current usage. This is not just a cultural historicity, but a psychological historicity, tied up with the problem of associations, the process by which perception and conceptualization link to develop significance with respect to form. Not surprisingly, Lerup sees the anthropomorphic trace everywhere in architecture, describing Eisenman's House X as a "plane of memory" in which a "trace of the body still appears as doors." This raises questions about whether such an inscription in architecture can bypass language and become a kind of writing, a making of forms that does not necessarily resort to explicit linguistic formulation. But at this juncture the question no longer involves simply comparing architecture to language as a symbolic system; it asks whether architecture is a form of presentation or representation. For the sake of discussing the formation of the subject, the link with language remains essential.

The discussion of the relations between language and architecture, as postulated by Eisenman and Lerup, forms the core of the concept of the subject. Even more important, it is the basis of making the distinction between the concept of the subject, in all its various forms, and that of the "person" against which it is being placed by both Lerup and Eisenman, and for which they each find an entirely different use. Eisenman limits his use of the comparison between architecture and language to a one-way modeling, letting architecture be analyzed and used as if it were a language. Eisenman's subject remains a linguistic user, cool and operational, the subject of a system without individual fears or desires. Lerup continually blurs the boundaries between language and architecture, using the constructions of one to challenge the codes of the other, tossing linguistic notions into architectural form and vice versa. His subject is the complex holographic production of a "text" that is at once architectural and linguistic, and it is also the psychoanalytic subject suffering the difficulties of formulation through identification in and with language and architecture as a means of producing identity.

The concept of the subject, in any of its many forms, has been challenged as antihumanistic. The concept of the person and of the inclusion of consideration of the person in architectural design is posed in contradistinction as compellingly attentive to the needs and desires of specific individuals. Lerup and Eisenman both point out the extent to which the interrelations of architecture and human subjects demonstrate the presence of an ideologically determined and coded subject already present in and produced by architectural discourse. The extent to which the person can be liberated from this process in order to function as an individual depends to some extent on the capacity to recognize the limitations of such individuality. The conventional sociological notion of the person, entrenched in traditional humanism, is challenged by the recognition of the extent to which architecture necessarily produces a subject of its own discourse. The myth of the individual, in architecture as well as other realms of human expression, has been undermined by the study of cultural practices that demonstrate the extent to which the individual is culturally produced and ideologically determined. Finally, it seems that against either mythic extreme—of the free-willed, free-thinking *individual* or of the culturally determined *subject*—must be posed another possibility that is not exactly a compromise negotiated between the two, although it acknowledges aspects of each concept. With full recognition of the constraints operating upon the individual and limiting the very differentiation process by which she or he is defined, the investigation of the production of an *individual subject* calls attention to these processes, whether in architecture or in language.

For Eisenman invoking the concept of the subject allows an investigation of architecture as a functioning system with its own constraints and operations, its own rules and limitations, within which the architecture attempts to create an articulation that bears the trace of self-conscious reflexivity. For Lerup the concept of the subject permits an exploration of architectural practice as an extension of psychological activity, as a formative and interactive mode. In both cases, the use of the subject promotes an attention to the architectural discourse in its specific and particular aspects in a way that neither Lerup nor Eisenman could articulate without it as a point of reference. For each of them invoking a subject grounds their investigation in architecture as a formal practice whose formality resonates through multiple levels of individual activity. If the person and the implied social responsibility that attends to its conception seems absent from the work, it is because the idea of a user is reformulated as integral with architecture, an already there, partially produced subject, for which the architect must also be accountable.

Notes

1. Mario Gandelsonas, "From Structure to Subject: The Formation of an Architectural Language," in Peter Eisenman, *House X* (New York: Rizzoli, 1982).
2. Ibid., p. 14.
3. Lars Lerup, *"House* X, Peter Eisenman," *Design Book Review* 2 (Summer 1983).
4. For extensive discussion and illustration of the Nofamily House, see Lars Lerup, *Planned Assaults* (Montreal and Cambridge: Canadian Center for Architecture and MIT Press, 1987), 20–58.

Architecture and Human Identity

JOSEPH B. JUHASZ

Issues of human identity are central to architecture if we accept the commonly held position that there are universal qualities defining good, beautiful, or attractive buildings. Whether Alexander's Pattern Language (1977), Charles Moore's psychologically based aesthetics (1980), or indeed Vitruvian principles of design, design principles assume a generalized human being (or generalized classes of human beings) about whom statements can be made irrespective of cultural milieu or specific individual identity. This generalized human being usually includes the architect, certainly includes the clients and users, and can in a metaphoric sense refer to the building itself. This chapter develops some principles for understanding what constitutes human nature as it pertains to architecture.

Who Is Human?

It is important to understand that who constitutes a human being is not itself a context-free, a culture-free or an ahistorical phenomenon. Definitions of humanity vary from place to place, time to time, and person to person, and they vary within the life span of a given person.

As we examine the shifts in definition of *human being* we note that the "edges" of humanity are open to negotiation. The borderlines are continually tested and explored (e.g., Is a fetus a human being?). This area of uncertainty in defining *human* reinforces an equally important fact about human identity: underlying any specific definition of *human* to be found in any given historical context or stage of the life cycle is a center that is all the more solid because its edges are blurred and shifting. The center of the concept is timeless, culture-free, and nonnegotiable: It is the biological[1] underpinning of human identity. Human identity indicates that we are members of the same species; we are offspring of others like ourselves.

In a similar way, although styles come and go, people can and do appreciate certain "timeless" structures and ways of building. The edges of these judgments are open to debate and redefinition. (Did the 1939 New York World's Fair provide an example of great architecture?) Yet the Pyramids, the Acropolis, and indeed that superficially different but structurally and thematically similar "special place" we all cherish are rarely questioned as touching a generalized sense of beauty.

Is There a "Human Nature"?

At least two extremely influential traditions in our time deny the reality of human nature. On the one side, existential theories assert that human nature is simply what we say it is. From this perspective human nature is an aspect of social identity. It is the most basic of our ascribed roles (Sarbin & Scheibe, 1980). This position is manifest in the work of nihilistic postmodern architects such as Graves or Venturi. Their architecture bespeaks the view that the user is what the user is defined to be, and that this is an unconstrained process. Much contemporary psychoarchitectural theory takes this perspective as its point of departure (e.g., Korosec-Serfaty, 1985; Dovey, 1985).

On the other side, "biological" theories attempt to divide the human race into meaningful subspecies. Coon (1962) asserted separate evolution of the main races of humanity, a claimed biological estrangement among the races that has a long history before and after his influential work. Architects who view their work as incomprehensible to those from other cultures (e.g., those who work totally "in a style," such as the present-day Johnson or Pei) fall within this category.

Existentialist Traditions

The root of the existentialist view is the belief that human beings are free to choose their "essence" (Sartre, 1956). This roughly means that the horizons of what it means to be human are only limited by our imagination. Frankly utopian schemes, such as those of the progenitors of the City Beautiful movement, Auroville, or Soleri's Arcology, presuppose a "human being" who must be encouraged to develop. The "architect as existential hero" stories (e.g., Rand, 1971) also have the theme that the creative artist works on a dramatically different horizon from the mere mortal. "Enlightened" architecture, such as the Alhambra, Sufi architecture, southwestern Kivas, or some traditional Taoist gardens, has a theme of liberation from the coils of fallen human nature through spiritual rebirth.

Existentialism was a reaction to beliefs about a totally fixed human nature. Such beliefs characterize traditional, feudalistic, preindustrial societies. In these societies the majority of social roles are ascribed, the majority of statuses are conferred for life. The belief in an utterly fixed human nature typically arises in societies in which sentiments, character, temperament, or traits can be seen as unchanging entities that reside in a people's ancestry—their "breeding." Nostalgic attempts to re-create such a real or imagined past often lie behind architectural "revivals" (e.g., Pugin, the arts-and-crafts school, the current fad for renovating "grand landmarks" such as the Willard Hotel in Washington, D.C.).

The changes from feudal to industrial underpinnings wreak havoc on the concept of fixed sentiments, character, temperament, and traits. As psychological trait theories are rejected in industrial society, it is natural to sweep away with them their conceptual superordinate—the belief in a fixed human nature. The citizen and the architect are now both faced with the fateful dilemma of seeming to be "free to contemplate their own destiny" (i.e., to experience existential anguish). The dissolution of the concept of a fixed personality is in fact associated with the "contradiction" between destiny and freedom—the contradiction some see fundamental to the tensions between planning and architecture that characterize the modern city (Tafuri, 1980).

The existentialist analysis categorizes human beings into two classes: the ignorant and slave and the learned and free. The ignorant slaves are products of their culture—immersed in it—rationalizers rather than rational. The few who are free and knowing, those belonging to the special culture of the culture-free social scientist or the artist-philosopher, transcend their ready-made social identities and are free to forge their own definitions of who they are. They are able to reason without having to rationalize.

The power of this view of human nature (i.e., that human nature is free to be defined as anything by the Hero who can accept the truth that He is without a Nature) is easy to see in architecture. This view was popularized by Ayn Rand's fictionalized portrayal of Frank Lloyd Wright as the existential hero. The view that the average human is "below" the architect pervades all the different strains and substrains of "modernism" and may almost be a definition of it. It is, especially in its noblest expressions (i.e., Mies), a patronizing attitude.

The Behaviorist Parallel

The behaviorist position represents an interesting parallel to that of the existentialists. Like the existentialists, the behaviorists represent the behavioral scientist–artist–visionary as the *ne plus ultra* exception to the rule of general ignorance, superstition, and rationalizing. Like the existentialists as well, they hold that for the average person there is no "human nature." Such an individual is a pawn of the environment.

The most extreme examples of fascist and protofascist architecture—whether from Stalinism, Hitlerism, Mussoliniism, Peronism, or Corbusierism—fall into this tradition. The Marseilles Block, Brasília, Foro Mussolini, or the Alexander Platz give such a totalitarian view of human nature.

In the case of the "behaviorist" artist and architect, the black knowledge that the machine age is the "Death of Art" obscures the reality from the "masses" just as much (Socialist-Realist kitsch) as the "existentialist" position of being "free to choose one's freedom" (Kahn). In either instance one is dealing with a conception of the artist as fundamentally different and removed from the ordinary run of humanity. Finally, in the blackest of ironies, the behaviorist takes a modified "free will" position too—for the elite behaviorist is in a position to know that he must "willingly submit to fate" (e.g., Le Corbusier, 1946).

These two relativist positions—existentialism and behaviorism—are sad monuments to the downfall of reason in the late twentieth century. Foucault (1972) has argued that the concept of "human being" is a fabrication of the Age of Reason—an outgrowth of the concept of reason itself. With the total victory of reason, who or what is human paradoxically became, for certain members of the intellectual–artistic elite, a question no longer possible to answer, for they no longer saw a thread that connects all humans.

With the victory of the industrial age, as in some of Sant Elia's most powerful renderings, the human element becomes invisible in the City. Relativism, which was a natural outgrowth of the expansion of the sphere of achieved roles in an industrial society, ultimately led to excessive claims about the malleability of human nature.

Biologically Based Theories

Because racism has been consciously repressed, its tentacles run even deeper than the still consciously articulated superrelativism of the existentialists and behaviorists. Biologically based theories deny the common humanity of the race. These theories identify different lines of descent for present social, racial, national, kinship, or other social groups.

European colonial architecture—whether the White-City-exported or earlier Hispanic, English, or French models—spoke in a direct way of the irreconcilability of the races. This was true of town planning as well as interior layout; it was even stylistically true. For this was not the liberal-minded exportation of the relativist, but an absolutism bespeaking the utter superiority of traditional European styles (Lerup, 1979). There was no deliberate attempt at hybridization—though, of course, unconscious borrowings abounded. Overtly, the architecture denied any kinship with those whom the buildings, cities, railroads, and kitchen plans oppressed; covertly, it was in fact spokesperson for that very kinship.

When investigators looked in areas populated by whites, they found parallel differences. These tendencies begot their architectural equivalent in the "folk styles" found throughout Europe between the wars. The neo-Teutonism of Hitler's architects, the nationalistic architectures of Eastern Europe (e.g., Lechner in Hungary), and the "Catalan Style" were all manifestations of the translation of colonialist concepts back in the Motherland.

The classifications of the races of the world and of Europe achieved great scientific, popular, and political currency. American immigration policies were changed in part through Madison Grant's (1916) version of the tripartite classification of Europeans. The ethnic "slum" and the "urban villager," on the one hand, and the American Prairie Style or Usonia, on the other, were architectural manifestations of this tendency.

These efforts, in effect, depended on evolutionary versions of a human family tree that assigned separate lines of descent to the races of the human species. The new historicism and eclecticism of the contemporary Graves, Johnson, and Pei seem concerned with similar considerations.[2] After all, an extreme historicism can be meant to say that if you do not share my history, if you cannot understand my position, then we lack a common fundamental vocabulary in which to converse. At its extreme this would be like saying that there is no point in attempting to create an "international style," for the kinship of humans is an illusion.

I categorized the relativism of the existentialist–behaviorist theories as reflections of mobile societies. From such a perspective the biologically based theories could be construed as reactions against mobility. They may be fueled by fear of interaction with strangers.

Certainly the concept of "defensible space" or the de facto fortressification of much of contemporary architecture—not only of Johnson and Pei but also of the brutalists such as Rudolph or the great mall-and-interior-courtyard builders such as Portman—bespeaks a paranoia about public spaces in which people can really fraternize. Instead, we have artificial creations of pseudo-public space that deny the kinship of different segments of society—a kind of spatial horror of "racemixing" in the Renaissance Centers, Helmsley Palaces, Trump Towers, and Watertower Places of contemporary urban agglomerations.

But that very horror of intermarriage, of miscegenation, underlying the racialist theories gives the lie to their claim that there is no common humanity that unites all of us. From the biological viewpoint the fatal attraction of the "lower" races for the members of the "higher" races is the proof of the long-term viability of such supposed "cross breeding." This is analogous to retrogressive utopias about vernacular folk architecture as being more viable and powerful than "effete," "high-style" architecture (Moore). If the traditional European–American culture is superior, then the Renaissance Center has really nothing to fear of Detroit, nor does Citicorp have anything to fear of Harlem. Architecturally they bespeak a fear not only of contamination from but of victory by the ghetto.

Human Nature and Architecture: Theme, Structure, and Identity

Having rejected the concept that there is no human nature, we can begin to sketch out what human nature is and how it is related to architecture. Who *are* architects' people? In daring to answer such a question we need to make explicit an assumption so far implicit in what we have said. Human nature is immanent in human works. It is the traces that people make that give the best clue of who they are.

We must first distinguish between three levels on which human nature operates: *theme, structure,* and *identity*. These three levels also determine the meaning, context, and effect of human creations. The artefacts with which we refashion the planet in our own image also have theme, structure, and identity.

Themes operate on the level of universals of human experience. Human beings build their life stories around themes. That is their nature. In building such stories they externalize some given moment in an artefact. Any architectural work will take a position on dimensions such as Dionysian–Apollonian, sky–earth, fire–water, ancestor–source, concrete–abstract, representation–presentation, conjunctive–disjunctive, outlining–penumbration. The theme relates the architectural work to the realm of experience, to the phenomenal, the empirical. Theme establishes the protagonist–antagonist relationship and defines the beginning, middle, and end of the story told by the building. A theme is a plot outline, a calling to action; it establishes a mood and defines the emotions, the symbols, the "animals" that will be called forth by the building. On a literal level, theme defines style and building type.

The structure of a life story, its order, is a manifestation of human intelligence. Human nature structures human perception, imagination, and action. This aspect of human nature is not a product of experience; it is the portentous "dream" that the species "dreams." It is the species' version of Order.

Architectural structure enframes the building in terms of the acceptance of premises, of the "willing suspension of disbelief" demanded by the edifice. The structure of the story is neither its plot, its symbolism, nor its main characters. Rather, structure relates to the portentous devices that establish the bases of fictions, stories, narratives, and theories. Structure antedates theme as dream antedates experience and as image antedates word and concept. Structural elements are those such as The Figure, The Shade, The Line, The Vortex, The Fold, The Crack, The Curve. There can be no themes without structure; there are structures without theme. Structure makes shape, order, contrast, brightness, volume, and loudness possible. Architectural structures operate on the levels of universals of dream and fantasy. On a literal level, structure keeps the building from falling down.

Identity relates theme and structure to person. Identity is relative, particular, and time bound. On the level of human nature, identity relates individual to society. Identity also relates architecture to society. The designer, the builder, the user, the restorer, the renovator, the decorator, the building inspector, the codes enforcer, the graffitoist cast a piece of their particular story into and off onto the building. On the literal level, identity establishes the period, the epoch, the history, the signature.

The building itself is a nest made from the excrescences of all who have handled it, seen it, worked on it, dreamt it. It represents a *position* on a collective idea and an act of theme, structure, and identity. As such it is neither "self-expression" nor "living object," "coin," "symbol," or "statement." Rather, it is a function of the state of its makers and users, of their dreams as well as of their experiences, of their social statuses and roles as well as of their interpretations.

In the remaining portion of this essay we briefly sketch the role of themes in human nature and architecture. A detailed exposition of the relation of theme to structure and identity awaits the time, energy, and intelligence of the author.

Themes in Architecture and Human Identity[3]

Thematically, human nature is a series of potentials within which a life story can be told. The buildings we fashion have the same potentials. They are, however, relatively more static than any human being, but not more static than an epoch, a social order, or an economic system. The themes are dimensional. Each human being has a "home" within that dimension. The life story is the story of home and exploration within each of the dimensions. Each building, too, has a home and a place within the dimensions. Its use and alteration is its "life story." The following list is sufficient to exhaust the list of themes, though none of the particular dimensions is necessary.

Dionysian–Apollonian. The Dionysian–Apollonian dimension occupies a space between the rational and ordered or the orgiastic and emotional.

Stylistically, this is the basis of the romantic–classicist dichotomy. Is the final building thematically rational (Mies) or orgiastic (Moore)? And what of the architect? The client? The building itself is not a self-portrait or a complete theory but the test of a position (e.g., architects like Wright and Corbusier were able to take numerous positions on this continuum with their buildings).

Sky–Earth. To be human is to have one's home in the sky or on the earth, and to be able to experience the polarity of being attached and attracted to those two homes. Does the user need to be elevated (Mies)? Reconnected to the ground (Goff)? Mediate between earth and sky (Wright)?

Fire–Water. Fire is said to be humanity's first and greatest invention—the use of fire is often used as hallmark of the species. It was "stolen from the gods." The union of water and earth forms clay—the archetypal stuff of cities, pots, bricks, idols, and Adam. Fire and water as emblems of Adam and Eve—our sources of invention and of life—show us another range of possibilities open to us as humans. Water and fire, most important, serve to unite earth and air. Home is the line between well and hearth. Gothic and Baroque architecture live in a fire theme. The classic architecture of Japan is a water architecture. Wright (e.g., the Guggenheim) and American pueblos (e.g., Taos) are examples of the clay-potter's balance.

Ancestor–Source. Adam is an ancestor; it is possible, even in a matrilineal society, to trace a line of descent to him. Eve, on the other hand, is the direct source of life—even in extremely patriarchal societies.

The source of funds is the client. The sources of the building are its builders. The ancestor is the architect. Often the users are perceived as unruly children of this complex union. The contemporary term for the babying of users is to say that a building does not call forth "environmental competence." Is it not the height of irony that the principal visible thrust for such competence is in the design of children's playgrounds (e.g., "adventure playgrounds")? Meanwhile, much of psychology's supposed contribution to design continues to be the call for environments that "fit" the user, that serve as no challenge—intellectual, emotional, or physical.

The species of ancestor worship exhibited in classicism has to do with the deifying of the indirect, as in the best work of Mies. In contrast, directly "feminine" forms, as in Gaudi's best work, worship sourceness. As a third reference case, some of the arid aristocracy of Wright's best work distances the user from the propagation of life. Here detached celibacy is deified.

Concrete–Abstract. The distinction between concrete and abstract deals with our ability and tendency to infer. The greater the amount of inference in a story, the stronger its abstract rather than concrete theme. People, perceived from the architect's stance, can be viewed as more or less concrete or abstract. Much of the constructed environment is "signed" with concrete (imaginal) or abstract (verbal) users in mind. Oddly, people-like-self are often targeted with verbal messages and people-unlike-self are targeted with imaginal messages. Here people-like-self refers to "standard users" who are assumed to be verbal. "Architects" are assumed to be not like self, in that they think imaginally.

The iconic means of communication often appears "inferior"; the verbal sign is the last word. It is always interesting to see instances where the architecture (which is in essence concrete) clashes with the abstract signs hung on it.

Representation–Presentation. Representation–presentation is a distinction between the talker and the doer. Directness in this sense is doing; it is presentational. And being indirect is associated with talking; it is representational. Many human activities and products are representational in that they refer to that which is known and articulated. Many other human activities and products do not refer to anything. They are or become, but they do not refer. Such activities, processes, or products are presentational.

Though it is perhaps too early to attempt to understand a Rossi, Graves, or Venturi completely, it is manifestly clear that what they engage in is to present and confront users with a confusing, rich, puzzling reality. Even Moore's Piazza d'Italia and World's Fair Wonderwall, in all their staginess and lack of shelter, speak to those that are present (reemanations of Romans, though they may be). The ghostly world and the real world have to collide directly in the architectural world. Architecture always has to have some manifest use along with the latent uses. The latent uses of architecture are fused with the manifest uses through the fiery oven of the design process.

Here again the distinction is between the verbally and the directly active person. The architect is a nonverbal actor in a world of verbal players. The disjunction between self and other (the hereness and thereness—the I am here and you are there, and vice versa—of the architect as against the client and users) is very strongly manifested here. The current stress on surface representation in architecture may be an attempt to heal this conflict. Alexander's Linz Cafe surely sees Man as Talker. The fantastic architecture of Sant Elia and the brutality of Rudolph celebrate the death of talk and canonize action.

Disjunctive–Conjunctive. There is a kind of "logic" that works by a principle of "either-or," a logic in which the principle of the excluded middle holds. There is another kind of "logic" that operates on the principle of "both/and." Here there is no law of contradiction. The former of these could be called a disjunctive and the latter a conjunctive principle. Human beings perceive themselves in a world of possibilities between adhering to one or another of these principles.

Either we are on the inside or we are on the outside, but not both—though we may play with the distinction (Csikszentmihalyi & Rochberg-Halton, 1981). This is primarily disjunctive. Yet the very act of separation joins the inside with the outside—glues them together. Thus, the building in affirming the either/or demonstrates the both/and.

Outlining–Penumbration. Outlining–penumbration speaks to two distinct ways in which form and thus story may be generated. Any form can be established by defining the outline of the edges of the form, or it can be established by showing how it reflects and absorbs light, by shading, gradation, chromaticity, movement, rhythm, and tempo.

Forming an object by penumbration is an attempt to render the "immediate" play of light and dark, reflection and shade. Whereas penumbration strives for the direct "sensation" of light and dark, outlining strives for the achievement of "perception" through the separation and disjoining of object and space.

In defining the edges we give boundaries. These boundaries give plasticity to the whole and set it apart from the empty space surrounding it. This gives us figure and ground, object and void. If we think of the object as a reflector, or source of light, we sculpt not the "object" but the void itself. This too establishes the dialogue between object and space, self and other—but the concern is for the center, not the boundary—for the materiality, not the surround.

The human identity characteristics that parallel the preceding approaches to making things are contrasting ways of distinguishing between the self and others. Adam sculpts, whittles with a knife. Eve plays with the elusive, equivocal brewing of light and shade, hill and valley, plain and defile.

These two modes of creation are *ex materia* and *ex nihilo*. We create ourselves—as well as the forms around us—by outlining (subtraction) and by penumbration (congealment). These two ways can be both antagonistic and complementary. In a sense they refer to the interplay of fantasy with physical form in the creation of our surroundings (Juhasz, 1976).

Unlike those mentioned earlier, this dimension of human nature is clearly "architectural." It is both least and most in touch with word-becoming-flesh.

Architects tend to perceive others as down to earth and as neither penumbrators nor outliners. They would like to perceive themselves as penumbrators, but know that mostly they hardline. Habraken (1972) is an architect–theoretician who sees people as outliners. Lucien Kroll (1980) sees people as penumbrators.

Architects' People

Architects' people are themselves. It is only the errors of arrogance, elitism, racism, relativism, projection, or narcissism that twist the truth of that statement into fallacy.

I teach a course called Techniques for Improving Imagining Ability. I try to teach prospective architects about themselves and about things. What I have discovered is that the more I can encourage my students to make things that are direct, immediate, personal, and identified, the easier it is for others to understand, react and respond to, or interpret the made object. This is not self-expression. Rather, it is the clarification of the themes of their life and the structure of their attachments. A particular human being fashions a particular object, but the story is general in its particularity; it can be used.

Not all things made by human beings are pretty or even beautiful or good, for human nature has its seamy, ugly, and evil aspects. All structure and all theme has potential attractiveness, beauty, goodness. The sources of goodness, beauty, and attraction lie not in the theme or structure but in the personal, in identity.

The Pryamids, the Acropolis, and indeed that superficially different but structurally and thematically similar "special place" we all cherish are not monuments to a particular ego. Nor are they egoless. Through personal and social identity they are a path into the themes and structures that fashion us as human beings. Great works of architecture are monuments to human nature—to the future of the species immanent in and portended by its past. Great works of architecture dream the human story in clay, brick, water—not in flesh, blood, and bone. But they are sensible signs of the invisible inner nature of those.

The empirical world is not ordered. Our sense of order does not derive from our observation of the world. Order is dreamt; it is immanent in the dream. Structure itself, order, does not come from the world, yet it is manifest in it. In this sense the externalizing of structure through theme and identity makes of the designer, the builder, the user, the restorer, the renovator, the decorator, the building inspector, the codes enforcer, the graffitoist—the rabbi, and the mohel—working heroes. Their red badge of courage is a nest-wound-scar-brand made from the excrescences of all who have handled it, seen it, worked on it, dreamt it. It represents a position on a collective idea and act of theme, structure, and identity.

Notes

I am deeply grateful to Russ Ellis, Ted Sarbin, Wendy Davis and Larry Goldberg for their help in the preparation of this manuscript. Ann Ellis, my collaborator on another work has also contributed a great deal to my thinking on this subject.

1. Biological in the most general sense of that word; meaning merely that it is the physical reality, the stuff, the is-ness and such-ness of being human.
2. For all that they would justly deny any intentional racism.
3. A more extensive—and earlier—version of these themes can be found in Juhasz (1983). In that version the themes are related to mythological stories of anthropogenesis.

References

Alexander, C., Ishikawa, S., and Silverstein, M. A Pattern Language. New York: Oxford University Press, 1977.

Coon, C. The Origin of the Races. New York: Alfred A. Knopf, 1962.

Csikszentmihalyi, M., and Rochberg-Halton, E. The Meaning of Things: Domestic Symbols and the Self. New York: Cambridge University Press, 1981.

Dovey, K. "Home and Homelessness." In Home Environments, edited by I. Altman & C. M. Werner, pp. 33–64. New York: Plenum, 1985.

Foucault, M. The Archeology of Knowledge. New York: Pantheon, 1972.

Grant, M. The Passing of the Great Race. New York: Scribner, 1916.

Habraken, N. J. Supports: Alternatives to Mass Housing. New York: Praeger, 1972.

Juhasz, J. B. Psychology and Physical Form. Berkeley, Calif.: College of Environmental Design, 1976.

Juhasz, J. B. "Social Identity in the Context of Human and Personal Identity." In Studies in Social Identity, edited by T. R. Sarbin and K. E. Scheibe, pp. 289–318. New York: Praeger, 1983.

Kroll, L. "Architecture and Bureaucracy." In Architecture for People, edited by B. Mikellides, pp. 162–70. New York: Holt, 1980.

Korosec-Serfaty, P. "Experience and Use of the Dwelling." In Home Environments, edited by I. Altman and C. M. Werner, pp. 65–86. New York: Plenum, 1985.

Le Corbusier, E. Towards a New Architecture. London: Architectural Press, 1946.

Lerup, L. Building the Unfinished: Architecture and Human Action. Beverly Hills, Calif.: Sage Publications, 1979.

Moore, C. "Human Energy." In Architecture for People, edited by B. M. Mikellides, pp. 115–21. New York: Holt Rinehart & Winston, 1980.

Rand, A. *The Fountainhead*. New York: New American Library, 1971.

Sarbin, T. R., and Scheibe, K. E. "The Transvaluation of Social Identity," In *The Normative Dimension in Public Administration*, edited by C. J. Bellone. New York: Marcel Dekker, 1980.

Sartre, J. P. *Being and Nothingness*. New York: Philosophical Library, 1956.

Tafuri, M. *Theories and History of Architecture*. New York: Harper & Row, 1980.

10

Authors and Architects

RAYMOND LIFCHEZ

Authors

Authors and architects have in common their desire to create buildings that work well. Each wants to place his clients or characters into settings that satisfy their physical and psychological needs and that are symbolically appropriate.

Architects are trained to consider human accommodation as a major factor in the design of buildings, and the credibility of an author's story rests largely on the resolution of characters and settings in mutually enhancing relationships. This way of thinking about the man–environment relationship is modern, as of the later eighteenth century, and continues to develop. A concept of man in the world as a psychological rather than purely spiritual being is a gift of the Western Enlightenment that gained increasing acceptance through the nineteenth century, a consequence of evolutionary thinking, resulting in the invention of the social sciences. In light of these advancements, the view of the role of architecture in shaping human affairs also changed. One consequence of the new concept of man in the world was a change in the perception of the role of architecture in human affairs. A new and complex dimension was given to the ancient uses of architectural symbolism.

The instrumental use of symbolism in architectural design is very old. The ancients were well versed in how to invest buildings with a particular social or creedal meaning, to express the invisible or intangible by means of visible or sensuous representations and thereby promote the integration of the populace into a received view of the world. As the shift in thinking about man in the world occurred, so too did a shift in the way architecture was considered. The conventional uses of symbolism gave way to a far more complex acceptance of the role of architecture in human affairs:

196

Architecture as a spatial creation is the outer garment of a secretive and vital system; it is a nonverbal manifestation of a preconscious condition. A completed and relatively fixed architectural structure is nevertheless a dynamic and organic entity, a system of coordinates that relates inner and outer spheres and in so doing creates a complex of new harmonies and tensions. Within its walls, columns, ceilings, chimneys, windows, turrets, or other structural elements, an edifice may be looked upon as a world in itself—a microcosm—an expression of a preexistent form that may be apprehended on a personal and temporal as well as a transpersonal and atemporal level.[1]

Eighteenth- and nineteenth-century authors understood architecture as having this potential and were early experimenters with architectural designs that were intended to touch the deeper structures of psychological life. In British literature we find a new architectural dimension in Gothic poetry and fiction, in which medieval architecture was used provocatively to turn the viewer inward, into a confrontation with the dark side of his being.

In the nineteenth century the psychological dimension was again broadened by the uses of metaphor in which a building and space were projected as the physical analogue of an inhabitant's personality and his role in his world. This mode is explored by Mary Shelley in *Frankenstein* (1831), in which the emotive intent of Gothic writing—with its reliance upon the chill induced by confrontation with the remnants of medieval architecture, and its "mystically tenanted chapel" and "activating priest"[2]—is given a modern architectural setting of the scientific laboratory where a human monster is created. Laboratory and scientist serve as metaphors for man's usurpation of Divine creation.

Robert Louis Stevenson, in *Dr. Jekyll and Mr. Hyde* (1885), uses architectural metaphor in yet a more complex way when he designs two completely different realms within the same house, each suited to that aspect of the double character who is to dwell within.

> Jekyll's transformation implies a concentration of evil that already inhabited him rather than a complete metamorphosis. Jekyll is not pure good, and Hyde . . . is not pure evil, for just as parts of unacceptable Hyde dwell within acceptable Jekyll, so over Hyde hovers a halo of Jekyll, horrified at his worser half's iniquity.
>
> The relations of the two are typified by Jekyll's house, which is half Jekyll and half Hyde. As Utterson and his friend Enfield were taking a ramble one Sunday, they came to a bystreet in a busy quarter of London which, though small and what is called quiet, drove a thriving trade on weekdays. "Even on Sunday, when it veiled its more florid charms and lay comparatively empty of passage, the street shone out in contrast to its dingy neighbourhood, like a fire in a forest; and with its freshly painted shutters, well-polished brasses, and general cleanliness and gaiety of note, instantly caught and pleased the eye of the passenger.
>
> "Two doors from one corner, on the left hand going east, the line was broken by the entry of a court; and just at that point, a certain sinister block of building

thrust forward its gable on the street. It was two storeys high; showed no window, nothing but a door on the lower storey and a blind forehead of discoloured wall on the upper; and bore in every feature, the marks of prolonged and sordid negligence. The door, which was equipped with neither bell nor knocker, was blistered and distained. Tramps slouched into the recess and struck matches on the panels; children kept shop upon the steps; the schoolboy had tried his knife on the mouldings; and for close on a generation, no one had appeared to drive away these random visitors or to repair their ravages. . . . There are three windows looking on the court on the first floor; none below; the windows are always shut but they're clean. And there is a chimney which is generally smoking; so somebody must live there. And yet it's not so sure; for the buildings are so packed together about that court, that it's hard to say where one ends and another begins."

Around the corner from the bystreet there is a square of ancient, handsome houses, somewhat run to seed and cut up into flats and chambers. "One house, however, second from the corner, was still occupied entire; and at the door of this, which wore a great air of wealth and comfort," Utterson was to knock and inquire for his friend, Dr. Jekyll. Utterson knows that the door of the building through which Mr. Hyde had passed is the door to the old dissecting room of the surgeon who had owned the house before Dr. Jekyll bought it and that it is a part of the elegant house fronting on the square. . . . Just as Jekyll is a mixture of good and bad, so Jekyll's dwelling place is also a mixture, a very neat symbol, a very neat representation of the Jekyll and Hyde relationship.[3]

It seems entirely probable that the Swiss psychologist Carl Jung learned from writers such as Shelley and Stevenson. It was he who was to make the connection between psychology and building that has had the most influence upon how we today consider the relationship. Jung's seminal thoughts on the subject are explored by Suzanne H. Crowhurst Lennard in *Explorations in the Meaning of Architecture,* a volume of essays devoted to writers and the houses they have built for themselves.[4] Jung's house at Bollingen, built by himself and over a number of years, served to anchor him in the ongoing process of self-analysis. Jung's house was both mnemonic and cathartic: As an artifact it served his memory and as a building process it was heuristic, revealing a preconscious condition to which he had had no other access. Through ongoing involvement with the house, as its builder and resident, Jung claims to have come to terms with deepseated problems and to have discovered aspects of himself he had never before explored nor expressed. On the occasion of his wife's death, Jung realized that there was yet one more aspect of his personality to consider:

To put it in the language of the Bollingen house, I suddenly realized that the small central section which crouched so low, so hidden, was myself! I could no longer hide myself behind the "maternal" and the "spiritual" towers [earlier additions]. So in that same year, I added an upper story to this section, which represents myself, or my ego-personality . . . with that the building was complete.[5]

Architecture has different roles in fiction and nonfiction, and both offer instruction to the architect in how to consider the client's world in the design task before him. In nonfiction—biography, autobiography, travel—buildings and settings are to be observed as parts of an individual's history and to be considered for their effect upon the subject's cultural and social underpinnings. In fiction the designer is shown ways to think more psychologically, to consider function in terms of both overt and covert behavior, in terms of the individual acting alone and in relationship to others.[6] Authors of fiction rarely describe their architectural design process, but Umberto Eco does so. In *Postscript to "The Name of the Rose,"* he makes it clear that the creation of characters is inseparable from the creation of the settings in which he wishes to place them.

> The first year of work on my novel was devoted to the construction of the world. Long registers of all the books that could be found in a medieval library. Lists of names and personal data for many characters, a number of whom were then excluded from the story. In other words, I had to know who the rest of the monks were, those who do not appear in the book. It was not necessary for the reader to know them, but I had to know them. Whoever said that fiction must compete with the city directory? Perhaps it must also compete with the planning board. Therefore I conducted long architectural investigations, studying photographs and floor plans in the encyclopedia of architecture, to establish the arrangement of the abbey, the distances, even the number of steps in a spiral staircase. The film director Marco Ferreri once said to me that my dialogue is like a movie's because it lasts exactly the right length of time. It had to. When two of my characters spoke while walking from the refectory to the cloister, I wrote with the plan before my eyes; and when they reached their destination, they stopped talking.[7]

Eco indicates the importance of knowing his characters' lives thoroughly, even more about them than he may care to reveal to the reader. The authenticity of the settings he will create for them is dependent upon the depth of his knowledge of them as dynamic beings, active in the world, as three-dimensional agents in time and space, not as set pieces in a frozen tableau. Eco uses this knowledge to advantage; while he creates situations in which characters act out their roles, he observes them, uses his watch to time them, notices their singular distractions from the task they have been "handed" to perform. From this exercise, in which character, act, and setting are combined, Eco determines the credibility of the scenario he has in mind as a writer telling the story. By this process the characters assist their creator, indeed, in some curious way, actually tell the story to the author, who then records what he hears and observes. In literature this is a process not without risk, as John Fowles discovered in writing both *The Magus* and *The French Lieutenant's Woman,* stories in which characters become so real, so independent as to leave the author puzzled about their motives, their decisions, and in the case of the latter novel, even about how they prefer to terminate *their* story to the reader.

In the examples we have seen, buildings are used as direct expressions of the people associated with them. To this genre of metaphor Lennard assigns the term *isomorphic,* being of the same or of like form; the converse is the "complementary" metaphor, in which the building or space is created or chosen in order to balance an individual's inner experience.[8] For example, a compulsive and driven collector of *objets d'art* chooses to sleep in a monkish cell; a vertiginous woman makes a tower her home; an illiterate man lines his walls with learned books.

It is with this reality of human nature in mind that Hermann Hesse designs a house and rooms in which to unfold the story of Harry Haller, the Steppenwolf. Haller, a middle-aged man, becomes a lodger in a bourgeois house. One of the other lodgers, a younger man, takes an interest in him and seeks to know him better.

> I gradually got acquainted with this strange man. . . . He lay always very late in bed. Often he was not up much before noon and went across from his bedroom to his sitting-room in his dressing-gown. This sitting-room, a large and comfortable attic room with two windows, after a few days was not at all the same as when occupied by other tenants. It filled up; and as time went on it was always fuller. Pictures were hung on the walls, drawings tacked up—sometimes illustrations cut out from magazines and often changed . . . some brightly painted water-colours, which, as we discovered later, he had painted himself . . . photographs of a pretty young woman . . . a Siamese Buddha hung on the wall, to be replaced first by Michelangelo's "Night," then by a portrait of the Mahatma Gandhi. Books filled the large bookcase and lay everywhere else as well, on the table, on the pretty old bureau, on the sofa, on the chairs and all about on the floor, books with notes slipped into them which were continually changing. The books constantly increased, for besides bringing whole armfuls back with him from the libraries he was always getting parcels of them by post. The occupant of this room might well be a learned man; and to this the all-pervading smell of cigar-smoke might testify as well as the stumps and ash of cigars all about the room.[9]

The contrast between Haller's dissolute way of life within the confines of his personal environment and his deep appreciation of the most orderly household strikes the young man as peculiar, and also touches him, for he senses that Haller is a deeply conflicted person. After Haller has mysteriously departed, the truth is revealed, in papers he leaves behind for the younger man:

A wild longing for strong emotions and sensations seethes in me, a rage against this toneless, flat, normal and sterile life. . . .

In this plight then, I went down the steep stairs from my attic-cell among strangers, those smug and well-brushed stairs of a three-storey house, let as three flats to highly respectable families. I don't know how it comes about, but I, the homeless Steppenwolf, the solitary, the hater of life's petty conventions, always take up my quarters in just such houses as this. It is an old weakness of mine. I live neither in palatial houses nor in those of the humble poor, but instead and deliberately in these respectable and wearisome and spotless middle class homes, which smell of turpentine and soap and where there is a panic if you bang the door or come in with dirty shoes. The love of this atomsphere comes, no doubt, from the days of my childhood, and a secret yearning I have for something homelike drives me, though with little hope, to follow the same old stupid road. *Then again, I like the contrast between my lonely, loveless, hunted, and thoroughly disorderly existence and this middle-class family-life.* I like to breathe in on the stairs this odour of quiet and order, of cleanliness and respectable domesticity. There is something in it that touches me in spite of my hatred for all it stands for. I like to step across the threshold of my room and leave it suddenly behind; to see, instead, cigar-ash and wine-bottles among the heaped-up books and there is nothing but disorder and neglect; and where everything—books, manuscript, thoughts—is marked and saturated with the plight of lonely men, with the problem of existence and with the yearning after a new orientation for an age that has lost its bearings.[10]

And thus in *Steppenwolf* we are confronted by the limitations imposed on the architect to use architecture as a mirror of the maneuvers of the mind to design a suitable home for the self.

At the outset of Evelyn Waugh's *Brideshead Revisited,* Charles Ryder recalls his university days, when he was befriended by Sebastian Flyte. They take an afternoon's drive to view the Flyte family's great country seat, an ancestral mansion evocative of a world far beyond Ryder's own middle class. Ryder is overwhelmed by what he is shown:

We drove on and in the early afternoon came to our destination: *wrought-iron gates* and *twin, classical lodges* on a village green, *an avenue, more gates, open parkland,* a turn in the drive; and suddenly a new and *secret landscape* opened before us. We were at the head of a valley and below us, *half a mile distant, prone in the sunlight, grey and gold amid a screen of boskage,* shone *the dome and columns* of an *old* house. . . . Beyond the dome lay receding steps of water and round it, *guarding* and *hiding* it, stood the soft hills. . . . We drove round the front into a side court . . . and entered through the *fortress-like, stone-flagged, stone-vaulted* passages of the servants' quarters.[11]

Ryder's description of his first view of the house indicates his position as the outsider who wishes in: The house seems to him a fortress panoplied by long vistas, iron gates, porticoes and domes, all "designed" to monumentalize the world in which the aristocracy live and to emphasize his distance from that world. But Ryder does gain admittance and becomes an intimate friend of the Flyte family. Some twenty-five years after that first visit, and after an absence of a few years, Ryder again sees the house. Now his vocabulary indicates his more complex relationship to the house and family to whom it still belongs:

> Our camp lay along one gentle slope; opposite us the ground led, still unravished, to the neighbourly horizon, and between us flowed a stream—it was named the Bride and rose not two miles away at a farm called Bridesprings, *where we used sometimes to walk to tea*; it became a considerable river lower down before it joined the Avon—which had been dammed here to form three lakes, one no more than a wet slate among the reeds, but the others more spacious, reflecting the clouds and the mighty beeches at their margin. The woods were all of oak and beech, the oak grey and bare, the beech faintly dusted with green by the breaking buds; they made a simple, carefully designed pattern with the green glades and the wide green spaces—*Did the fallow deer graze here still?*—and, lest the eye wander aimlessly, a Doric temple stood by the water's edge, and an ivy-grown arch spanned the lowest of the connecting weirs. All this had been planned and planted a century and a half ago so that, at about this date, it might be seen in its maturity. *From where I stood the house was hidden by a green spur, but I knew well how and where it lay, couched among the lime trees like a hind in the bracken.* [12]

Novelist Waugh creates two different views of the same landscape in which to cast Ryder—one for the young outsider, the other for the older insider, each appropriate to the two stages in Ryder's life and to what the reader knows about Ryder at a given point in the story. Through the perceptions of the house, its natural and human landscape, not only the architecture of the story, but the personal relationships are developed and revealed. The great house stands as a metaphor for the human drama played out within and becomes an architectural archetype of the house as matriarchate. [13]

To use the concept of isomorphic and complementary fit there must already exist a recognition of the client's uniqueness by the architect. In this regard, children, the elderly, and people with physical disabilities are especially problematic, as they are often obscured by misconceptions or fears of which the designer must be relieved before his vision clears. The architect can turn to literature to find models for clients and mount finely ground lenses through which to observe them. The child client is a case in point.

Nurseries, playgrounds, and schoolrooms have a definite place in the scheme of things, have specific normative or remedial goals, and may require special knowledge to design. But these are places in the environment conceived of by adults for children and may not be attractive to children or beneficial to their psychological development, unless the designer can be reminded of what adults tend to forget about the experience of being a child. As slim excerpts from Nabokov, Proust, and Sartre illustrate, literature can supply precisely what is needed.

When Nabokov was a child, his family lived each summer in a country house that, in *Speak, Memory*, he recalls in great detail as a setting imbued in his mind with the important nightly ritual of going to bed:

> I next see my mother leading me bedward through the enormous hall, where a central flight of stairs swept up and up, with nothing but hothouse-like panes of glass between the upper landing and the light green evening sky. One would lag back and shuffle and slide a little on the smooth stone floor of the hall, causing the gentle hand at the small of one's back to propel one's reluctant frame by means of indulgent pushes. Upon reaching the stairway, my custom was to get to the steps by squirming under the hardrail between the newel post and the first banister. With every new summer, the process of squeezing through became more difficult; nowadays, even my ghost would get stuck.
>
> Another part of the ritual was to ascend with closed eyes. "Step, step, step," came my mother's voice as she led me up—and sure enough, the surface of the next tread would receive the blind child's confident foot; all one had to do was lift it a little higher than usual, so as to avoid stubbing one's toe against the riser. This slow, somewhat somnambulistic ascension in self-engendered darkness held obvious delights. The keenest of them was not knowing when the last step would come. At the top of the stairs, one's foot would be automatically lifted to the deceptive call of "Step," and then, with a momentary sense of exquisite panic, with a wild contraction of muscles, would sink into the phantasm of a step, padded, as it were, with the infinitely elastic stuff of its own nonexistence.
>
> It is surprising what method there was in my bedtime dawdling. True, the whole going-up-the-stairs business now reveals certain transcendental values. Actually, however, I was merely playing for time by extending every second to its utmost.[14]

Nabokov's child-self reveals an exquisite fantasy reenacted nightly in relation to a very ordinary utilitarian object, the house stairway, involving a wondrous interpretation through which he asserts his identity while submitting to parental authority: an interpretation through which the child establishes ownership, making the stairway into what my colleague Peter Prangnell once termed a "friendly object." By incorporating the main household arterial way into his own territory, the boy asserts his identity within the adult world. This meaning establishes how the child sees the stairway and the way in which he is prepared to act toward it.

Jean-Paul Sartre's autobiography, *The Words*, gives yet another view of the child in the adult world. In this case, the child associates his attachment to an adult with that adult's personal space. The association facilitates his growing up and realizing himself as an individual; he separates himself from the other. Through this relationship he comes to understand relationships between adults, to understand the concept of work as an integral part of adult life, and thereby glimpse what he, too, might become as an adult.

> I began my life as I shall no doubt end it: amidst books. In my grandfather's study there were books everywhere. It was forbidden to dust them, except once a year, before the beginning of the October term. Though I did not yet know how to read, I already revered those standing stones: upright or leaning over, close together like bricks on the bookshelves or spaced out nobly in lanes of menhirs. I felt that our family's prosperity depended on them. They all looked alike. I disported myself in a tiny sanctuary, surrounded by ancient, heavy-set monuments which had seen me into the world, which would see me out of it, and whose permanence guaranteed me a future as calm as the past. I would touch them secretly to honor my hands with their dust, but I did not quite know what to do with them, and I was a daily witness of ceremonies whose meaning escaped me: my grandfather—who was usually so clumsy that my grandmother buttoned his gloves for him—handled those cultural objects with the dexterity of an officiant. Hundreds of times I saw him get up from his chair with an absent-minded look, walk around his table, cross the room in two strides, take down a volume without hesitating, without giving himself time to choose, leaf through it with a combined movement of his thumb and forefinger as he walked back to his chair, then, as soon as he was seated, open it sharply "to the right page," making it creak like a shoe. At times, I would draw near to observe those boxes which slit open like oysters, and I would see the nudity of their inner organs, pale, fusty leaves, slightly bloated, covered with black veinlets, which drank ink and smelled of mushrooms.[15]

Finally there is Marcel Proust, whose account of the madeleine has become synonymous with sensual effects and childhood pleasures. To trace Proust's childhood with him through *Remembrance of Things Past* is to remind oneself of the lessons and pleasures derived from the adult realm when a child feels himself to be an integral, and desired, part of it.

> They were rooms of that country order which . . . fascinate our sense of smell with the countless odours springing from their own special virtues, wisdom, habits, a whole secret system of life, invisible, superabundant and profoundly moral, which their atmosphere holds in solution; smells natural enough indeed, and coloured by circumstances as are those of the neighbouring countryside, but already humanised, domesticated, confined, an exquisite, skilful, limpid jelly, blending all the fruits of the season which have left the orchard for the store-room, smells changing with the year, but plenishing, domestic smells, which compensate for the sharpness of hoar frost with the sweet savour of warm bread, smells lazy and punctual as a village clock, roving smells, pious smells; rejoicing in a peace which brings only an increase of anxiety, and in a prosiness which serves as a deep source of poetry to the stranger who passes through their midst without having lived amongst them. The air of those rooms was saturated with the fine bouquet of a silence so nourishing, so succulent that I could not enter them without a sort of greedy enjoyment.[16]

The joys of childhood are thrown into high relief, however, when the child is also viewed as part of the ongoing world. Designing for a family or group with the children in mind inadvertently raises issues of peer and intergenerational conflict, struggles for personal identity and authority, and how these are enacted within and through the spatial organization of the house. When these kinds of issues emerge in the architect's deliberations, it is a good sign that his clients have achieved a kind of psychological maturity to him and are not just pro forma. Unscrambling of the thorny interpersonal problems draws greater attention to design solutions and their workability. These considerations also bring to light the sometimes difficult realization that the house is designed only once but the clients' lives are dynamic. The best the architect can hope for is that the house he turns over to them has a reasonable chance of being accommodating, at least in the foreseeable future. For exactly how long the house will be serviceable in predictable ways may depend upon the understanding achieved between client and architect as the house is being designed.

There are remarkable documentations of family lives lived within a single house. Virginia Woolf's essay "A Sketch of the Past" is one of the most provocative of these in its record of the ceaseless transformations of personalities and relationships played off within an unyielding Victorian regime entrenched in its authority by the house it occupies. Woolf is thirteen years old when her mother dies, and for the first time she is conscious of and affronted by Victorian charade, now of grief, in which she is expected to take part. This wrenching experience marks a passage in her life and is closely associated with the house itself, as she tells us in her description of a household in mourning:

> We in the front room sat crouched, hearing muffled voices, ready for the visitor to emerge [from the inner room] with tears on tear-stained cheeks. The shrouded, cautious, dulled life took the place of all the chatter and laughter of the summer. There were no more parties; no more young men and women laughing. No more flashing visions of white summer dresses and hansoms dashing off to private views and dinner parties, none of that natural life and gaiety which my mother had created. The grown-up world into which I would dash for a moment and pick off some joke or little scene and dash back again upstairs to the nursery was ended.[17]

Once recognized, the contradictions unleashed in her adolescence become more and more instrumental in shaping her image of herself. In middle age, she recalls her last years in her father's house; still a young woman, she and her sister Vanessa performed the duties of its mistress:

> Two different ages confronted each other in the drawing room at Hyde Park Gate: the Victorian age; and the Edwardian age. We were not his [their father's]

children, but his grandchildren. When we both felt that he was not only terrifying but also ridiculous. . . . The cruel thing was that while we could see the future, we were completely in the power of the past. That bred a violent struggle. By nature, both Vanessa and I were explorers, revolutionists, reformers. But our surroundings were at least fifty years behind the times. Father himself was a typical Victorian: George and Gerald [their older half-brothers] were unspeakably conventional. So that while we fought against them as individuals we also fought against them in their public capacity. We were living say in 1910: they were living in 1860.

In 22 Hyde Park Gate around about 1900 there was to be found a complete model of Victorian society. If I had the power to lift out a month of life as we lived it about 1900 I could extract a section of Victorian life, like one of those cases with glass covers in which one is shown ants or bees going about their affairs.[18]

Lest we draw the wrong conclusion about "A Sketch of the Past," it must be pointed out that it is also a story of childhood happinesses. The piece I like best deals with the bending of household rules and the complicity of the children's natural ally, the servants. Like Nabokov's scenario of going upstairs, Woolf's nighttime ritual involves an element of defiance that children must risk in order to discover how to manage the adult environment to their own ends.

The kitchen, Sophie's kitchen, was directly beneath our night nursery. We would let down a basket on a string and dangle it over the kitchen window, at night while dinner was going on. If she were in a good temper, the basket would be drawn in and laden with something left from the grown-ups' dinner; but if she were in a bad temper, the basket would be jerked in and the string cut. I can remember the different sensations: drawing up the heavy basket; and feeling the jerk; and the lightness of the string.[19]

These excerpts expose us to this family's life and some of its complexities, and remind architects of what is often overlooked: that conflict of various kinds is an unavoidable, even useful, part of everyday life and must be considered as a variable in each design problem. The family will arrange and rearrange the house, using its spaces and furnishing as "equipment" in the family's ongoing process of conflict resolution.

Architects

Aside from the various pleasures architects and their students derive from reading fiction and nonfiction, literature has several particular contributions to make to their professional development. First, literature can liberate us from the blinders of our social class, age group, and regional or national prejudices. As we read, the text opens for us a Pandora's box of images and emotions, some familiar and some exotic, oneiric, irrational, or fantastic. Second, in reading written descriptions of realistic and imaginary forms, we enhance our abilities to perceive and articulate perceptions through design, for literature presents new vistas that stimulate our imagination, sense organs, and emotive empathy. Third, each work of literature offers a new model of harmony, appropriateness, and beauty. Reading stimulates us to reexamine our concepts of form and function, our understanding of people, the natural world, the built environment, and the relations among them.

In beginning design courses, to establish a buildings-are-for-people theme, students are often asked to design a building for themselves. The next step is usually an exercise in which there is a "client" (other than the student) and the instructor "speaks" for him, deciding how well the student succeeds in making a suitable building. Later on, visiting professionals are introduced to bring a fresh and often more critical view of the student's work. In yet another approach, which I prefer, students invent their clients and considerable time is spent—as part of the design process—in developing the imagined clients' biographies and environmental preferences.

In this approach the clients, though ostensibly invented, are usually based on people the student knows or would like to know. Our aim is to lead the student into a professional relationship with the client—one that, as in the writing of fiction—will establish the client's independence. When a student counters a design recommendation by others with "my client would not like that!" we know that the student has succeeded in allowing the client's autonomous voice to enter the design process.

Generally, students have no difficulty creating clients. Their choices about their clients' age, gender, ethnicity, work, and other attributes tend to reflect something that may be alien or attractive in their own lives about which they are curious. But some students reject the exercise on the basis of architecture (as opposed to construction) as an art form; they protest that such practicality is demeaning. For these students, people are an intrusion into the architect's realm of creativity, and they subvert the exercise by inventing absentee clients—racy, multinational figures for whom the building in question is a "showcase" and whose only concerns are with architecture as art.

Other students are willing to try the exercise, but they produce stereotyped clients. The propensity to stereotype is also rooted in deep-seated views not readily dislodged about architecture and community. Such students produce conventional buildings, and this is disheartening when a student shows promise in other ways as a designer.

Most students come up with workable clients, interesting creations reflecting an inner need to integrate personal and academic pursuits. There is a strong identification of self with the client, and this makes the student more self-involved and leads to the best results. These evoked clients sometimes embody some part of the student that he wishes to acknowledge—be it troublesome or admirable; someone he identifies as a success or failure and with whom he is preoccupied (often a parent or parent figure); someone for whom he feels an emotional attachment that he wishes to examine or strengthen (deceased parents or grandparents, a contemporary friend or lover); some person or type with whom the student is uneasy or even fearful (the elderly, disabled, sexually permissive, or socially deviant); or sometimes an intriguing relationship the student has yet to initiate in his life (marriage, parenting, professional success). Sometimes students bring actual people together into a fictional client group, such as a student's own family, for whom he wants to design a house in which they will find contentment they now lack.

- A student chooses an elderly woman as her principal client. Later she reveals that the client is actually the grandmother she never knew. The design becomes a search for an important figure in her family's life.
- A student designs for a young woman, her husband, and his young children by a previous marriage. The student identifies herself in this relationship and puzzles out whether it is a viable one by setting it in the archetypal form of the home, to her the symbol of the family in normal society.
- A foreign student longs for his family and feels divided between life in America and life at home. He designs a house for his family in America to test out the feelings and practicality of superimposing one world upon the other.
- A student creates an interracial family and through the design of a house for them ponders her future with a fiancee of another culture.
- A student who has recently experienced his first homosexual relationship designs an apartment for a male couple and searches for an arrangement of spaces in which the couple will feel free of the intrusion of the values of the middle-class family in their life together.

Students introduce their clients with written biographies and short "scenarios," situations indicative of who the clients are, their relationships, their environmental preferences, and so on. The design itself is then started. As drawings and models develop, these are used as projections of how the student architect is "thinking" about the clients and how the clients are "thinking" about the design. Notions about function and behavior expressed in the designs are examined in the same terms as they might be in the analysis of a scenario drawn from literature. Concepts of symbolism and metaphor presented in works by authors in literature become useful examples by which to extend the fledging designer's thinking beyond the conventional. The direction a client's life has taken, which has led to the need for a new dwelling, may be examined in terms of his early life (Sartre in his grandfather's library). Students are encouraged to plumb the clients' environmental history to find a creative concept for the design of the house.

It is common for the beginning student of design, when thinking about the relationship between function and time, to isolate a frozen moment in time—a misconception that goes unnoticed unless the design is developed in terms of lives. The clients are essential to this aspect of the architectural discussion; without them the conventional signifiers of bathroom, kitchen, or living room are adequate. Once the human factor is introduced, time—momentary, diurnal, seasonal, generational—and its implications of change become central concerns. The concepts of need and preference enter as the notion of diverse roles (the insider or the outsider in *Brideshead Revisited*), how a setting is perceived, and the effect of perception on feelings and behavior are discussed. Interpersonal and group dynamics are essential determinants of how individuals use the environment to identify themselves, play a role, get what they need. This point is made well in autobiographies such as Virginia Woolf's. To examine these kinds of questions, more than one design solution to a particular setting may be required, from which essential elements might then be extracted and combined into a final program.

Like authors, architects must have a cast of characters in mind to obtain richness in design. But it is essential for design students to learn that there are serious limits on a designer's ability to anticipate how a design will be used. The client, no matter how well known, is but slightly known. Once a writer has completed a book, he or she finally surrenders the universe created therein to readers, who will in many ways make of the text what they will—to the author's delight or despair. So, too, a work of architecture will be continuously reinvented by the individuals who inhabit it.

Notes

1. Bettina Knapp, *Archetype, Architecture, and the Writer* (Bloomington: Indiana University Press, 1986), p. vi.
2. Kenneth Clark, *The Gothic Revival* (London: Constable, 1950), pp. 36–59. Also, George Levine and U. C. Knoepflmacher, *The Endurance of Frankenstein* (Berkeley: University of California Press, 1979), p. 36.
3. Vladimir Nabokov, *Lectures on Literature* (Orlando, Fla.: Harcourt Brace Jovanovich, 1980), pp. 184–86.

4. Suzanne H. Crowhurst Lennard, *Explorations in the Meaning of Architecture* (Woodstock, N.Y.: Gondolier Press, 1979).
5. Carl Jung, *Memories, Dreams, Reflections*, cited in Lennard, p. 25.
6. Herbert Blumer, *Symbolic Interactionism: Perspective and Method* (Englewood Cliffs, N.J.: Prentice-Hall, 1969). Also, R. Lifchez, "Being There: Intentionality in Architecture," *Journal of Architectural Education* 27 (June 1974): 42–51.
7. Umberto Eco, *Postscript to "The Name of the Rose,"* trans. William Weaver (Orlando, Fla.: Harcourt Brace Jovanovich, 1984), pp. 24–25.
8. Suzanne H. Crowhurst Lennard, "A House Is a Metaphor," *Journal of Architectural Education* 27 (June 1974): 35 ff.
9. Hermann Hesse, *Steppenwolf*, trans. Basil Creighton (New York: Frederick Unger, 1957), 7 ff.
10. Ibid., pp. 33–35; my emphasis.
11. Evelyn Waugh, *Brideshead Revisited: The Sacred and Profane Memories of Captain Charles Ryder* (1944; Boston: Little, Brown, 1973), p. 29; my emphasis.
12. Ibid., pp. 34–35; my emphasis.
13. Knapp, pp. 84–99.
14. Vladimir Nabokov, *Speak, Memory: An Autobiography Revisited* (New York: G. P. Putnam's Sons, 1966), pp. 83–84.
15. Jean-Paul Sartre, *The Words*, trans. Bernard Frechtman (New York: Braziller, 1964), pp. 40–41.
16. Marcel Proust, *Remembrance of Things Past*, trans. C. K. Scott Moncrieff (1913; New York: Random House, 1934), Vol. 1, p. 38.
17. Virginia Woolf, *Moments of Being* (Orlando, Fla.: Harcourt Brace Jovanovich, 1976), p. 94.
18. Ibid., pp. 126–27.
19. Ibid., p. 114.

IV
MISFUNCTIONS AND REVISIONS

The chapters thus far have rendered the architects' people from designers' words and buildings, as well as from texts such as myth and the novel. The construct of architects' people has been expanded by these studies, transforming any image of a diminutive internal person guiding the architect's hand into a dynamic, interactive, multivalent subject. It is now appropriate to reconsider the architect's image of the person in a larger social existence. Earlier chapters touched upon these larger patterns: Kahn's image of institutions, Wright's democracy, the New York architects' chaotic, ever-changing social fluid, Vitruvian class structure, and the Japanese culture's form without specified function. Although the architects' *person* has been linked to the designer's personal identity, the intersubjective image of others, and to the anthropomorphic building, the architects' *society* is more obscure.

The final chapters critically examine the architects' images of the widest possible social arena, entailing cultural, political, and economic forces, all of which help to shape people for architects and reciprocally, architecture for society. These forces are examined in terms of the policy-induced misfunction of what Groth calls "public architecture," the Post-Modernists' revision of society and Amendola's revisioning of the movement, and Montgomery's reexamination of misconceptions about functionalism and the nature of architectural work. These misfunctions and revisions fit within the critical perspective common to all chapters in this final section of the book.

Beginning the section is Groth's case study of "public architecture" as a tool of political action for social change. He traces one group of people, the single-room-occupancy tenants, and reveals how they were rendered invisible by architects and planners. This myopic social vision, by exclusion, defines the establishment view of the moral order. In the very first chapter, Favro explained that

Vituvius and Alberti refused to incorporate into their vision certain "mean" segments of society. These are Groth's "nonpeople."

Amendola next finds that the foundations of the current postmodern movement lie in a crisis over the architects' people. Whereas Groth puts forth the concept of nonpeople, Amendola's contemporary architects construe it as people-as-problem. He argues that the diversity of present-day society has threatened architecture's ability to communicate, thus promoting an architecture that disguises its political agenda. Robert Kerr's clear diagram of class structure and gender distinctions stands in stark contrast to the dissimulation Amendola describes. On the other hand, Wright's democratic utopia and the Post-Modernist camouflage of establishment values are both examples that architects and architecture are transmitters of a social vision. This drops the architects and their people directly into the political arena. Implicitly or explicitly guided by visions of utopia or maintenance of the status quo, architects make manifest their images of society in sticks and stones.

These chapters also bring the individual architect into the collective realm by considering the profession's sociohistorical position. The book closes with a revision by Montgomery of functionalism's ideology and subsequent developments within the profession that have created the current situation. Using Pruitt-Igoe as the key example, he demonstrates that changes in the production of the built environment have spurred the development of myths about the architect's role in society. Montgomery elaborates the new conditions of architectural work, suggesting that social scientists and architects worked together to invent architects' people. These inventions populate a revisionist form of modernism, but they are also the consumers of a highly compartmentalized design process that includes market research, programming, and evaluation.

The architectural project of making buildings inherently incorporates ideas about the person, group, and society. When the noted sociologist of art, Howard Becker, published *Art Worlds* (1983), he made clear that art is a collective activity. "Works of art . . . are not the products of individual makers, 'artists' who possess a rare and special gift. They are, rather, joint products of all the people who cooperate via an art world's characteristic conventions to brings works like that into existence." What we and Becker have overlooked is the participation of our perceptions of all those people. Although the patron, the beholder, the colleague, the institution, and the state have an existence outside of us, they also exist within, where perhaps they play an equally significant role in the creative architectural endeavor.

11

Nonpeople: A Case Study of Public Architects and Impaired Social Vision

PAUL GROTH

In 1910, several million Americans lived permanently in hotels. Twenty years later, planners and architects had erased those hotel residents from professional view and had planted one of the roots of the present homeless crisis. When an architect's view of humankind influences a small private building, the effects are not widespread; however, when a designer's definition of a human being influences hundreds or thousands of projects, the issue rises above academic or psychological curiosity to political and social importance.

Decisions or designs at the general or policy level are part of "public architecture"—the official and professional practice of building and managing architectural space, usually on behalf of public clients or governmental organizations. The people involved in public architecture make decisions not only about buildings or streets but also about building codes, planning rules, repeated and routine structures (as opposed to monumental public buildings), guidelines for public housing programs, or public funding or mortgage guarantees. Such practice involves designers as well as social workers, planners, lawyers, budget analysts, real estate experts, policy experts, and politicians.

Decision makers practice public architecture whenever they consciously use buildings or space as political tools for social change. These decision makers limit and rearrange space on behalf of the public even when the public does not directly participate. In effect, the bureaucrats who steer the codes and public funding for a building are "space police" who keep design behavior within both explicit and assumed boundaries established over decades of experiment, debate, or tradition.

In the practice of public architecture, individual limitations or blind spots can have far-reaching repercussions. This is shown very clearly in the his-

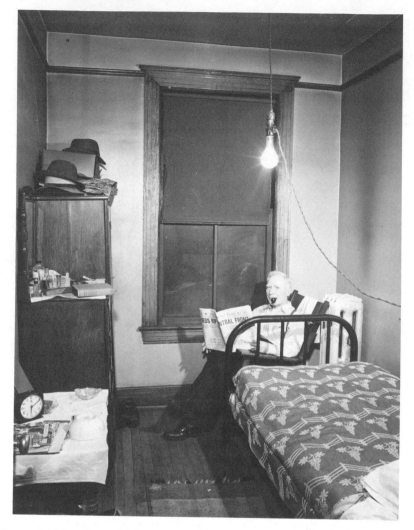

Figure 11.1 A long-term SRO tenant in an Indiana residential hotel, 1943. (Farm Security Administration, Library of Congress.)

tory of professional attitudes about the people who lived permanently in American hotel rooms. The historical residential hotel was much like its present-day survivors. In most states, *hotel* legally denoted a wide range of hotels, motels, rooming houses, or lodging houses—in fact, any building with six or more rooms rented for the day, week, or month that did not have private kitchens and baths (and would therefore qualify as apartments or tenements). "Residential" hotel rooms had tenants who had stayed for a month or more.[1] Residential hotels were 500-room public palaces or ten-room dives above a bar; rooms may have offered private baths or baths far down the hall. If kitchens were available, they were shared.

To the public architects who tore down and then partially rebuilt downtown America from the 1930s through the 1970s, hotel residents were invisible, even if they had lived there for fifteen years. Planners dubbed these people "SRO tenants" (SRO, meaning "Single Room Occupancy"; Figure 11.1). Because planners did not count SRO tenants as legal residents when renewal projects tore down their housing, in the official vision the action dislocated no one, even though thousands were moved.

Today between 1 and 2 million Americans still live in hotels—more than all the inhabitants of America's public housing. How did their far greater numbers at the turn of the century come to be invisible? How, in short, did they become *nonpeople* in official eyes?

This situation evolved in stages: first, seeing people negatively; second, filtering those people out of professional view while forming positive definitions of other public clients; third, looking away from the negative group while making public plans for the positive group; and finally, ignoring the negative people because they had ceased to exist officially. The following sections trace these stages, but first we must see hotel people as downtown entrepreneurs and employers saw them—as both normal and essential to urban life.

The Marketplace Vision: SRO Residents as Downtown Citizens

The tradition of using hotels as urban housing stretches back almost 200 years. Throughout the 1800s, hotels sheltered not only travelers but also long-term guests who needed a centrally located home where they could purchase all household chores and provisions as part of the rent.[2] Hotel keepers developed a sharp view of this large market and developed distinct building types for distinct social and economic classes of hotel patrons.

Wealthy people, many of whom had country or resort homes, often lived part of the year in palatial hotels. In 1907, for instance, the New York City papers published a cutaway diagram to show readers the new downtown homes of the Harrimans, Goulds, and Vanderbilts (Figure 11.2).[3] For about half the price in food and room costs, mid-priced hotels offered convenient household accommodations to middle-class politicians, writers, actors, business people, and professional women. At the least, people rented a single room; most couples rented a two-room suite. A hotel variant, the apartment hotel, offered larger suites and the regular use of an elegant public dining room.

Figure 11.2 A 1907 newspaper view of New York's new Plaza Hotel, showing its wealthiest tenants. (From Dorsey and Devine, 1964.)

The individual suite layouts in the Hotel Beresford, built in the 1880s on New York's fashionable Central Park West, display Victorian middle-class social conventions for hotel homes (Figure 11.3). Hallways were prominent and assured maximum separation of functions, as did the laby-

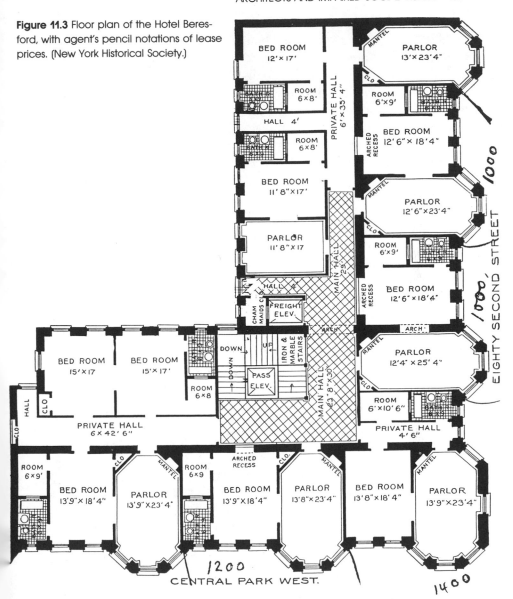

Figure 11.3 Floor plan of the Hotel Beresford, with agent's pencil notations of lease prices. (New York Historical Society.)

rinthian halls of New York apartments of that period. Individual units had no entry foyers, but the entry to the building served that function, and the public halls were decorated as in private homes. The Beresford's developers took into account the social standing of their clients by providing relatively large rooms (parlors averaging 14 by 23 feet), private baths, wide public stairs, an elevator, large air wells (for the time), and fire escapes *inside* the building (not, as with tenements, hung on the outside), in addition to the address, view, central dining room, lobby, and attentive hotel service. As the structures show, the architects of hotels such as the Plaza and the Beresford took their residents seriously and provided unique settings to attract them and to display their social positions.

Although single people predominated in all types of hotels, in the better hotels one would see many married couples and children. In 1920, the families of about one in twelve of Chicago's professional or mid-level business people lived in hotels equivalent to the Beresford. Calvin Coolidge and his wife, for instance, lived in hotels most of their lives.[4]

As late as 1929, hotel managers saw wealthy and middle-class people as an expanding market.[5] Financially secure residents could afford other housing, yet chose hotel convenience and lower costs. The vast majority of hotel tenants, however, had less economic choice.

About a third of a typical city's hotel homes catered to skilled workers who had relatively low pay, such as secretaries, department store clerks, junior salespeople, or skilled construction workers living in the city for a year or more during a building boom. This housing group was about half men and half women. Throughout the nineteenth century many families took in one or two of these people as boarders (for room and food) or lodgers (for room only). Alternatively, some entrepreneurs converted a formerly single-family house into a full-scale boarding house with six to fifteen tenants sharing a common bath. By 1900, most of the boarding-house operators in large cities had dropped the dining option and simply ran rooming houses.[6]

Rooming house residents lived in a social limbo. They were single and could barely afford to dress well. Rightly or not, the proper middle class scorned the districts where rooming houses prevailed. The rooming house *buildings* lived in limbo too: outwardly they were family structures, but inwardly, a series of individual rooms rented by people who did not know one another. The owners were poised to sell or convert the lot to some other use. Residents shared no parlor or dining room and had very different work schedules.

By about 1900, as owners rebuilt, they provided new, forthright structures as more permanent places for roomers (Figure 11.4). These downtown rooming houses revealed a centuries-old mixed-use scheme: stores on the first floor and a small door and stairs at the side, leading toward 12 to 35 rooms (about 10 feet square) on the upper floors (Figure 11.5).[7] Some operators built 200-room versions with the same plain but serviceable features.

Figure 11.4 Three downtown rooming houses, built ca. 1910 in San Francisco. The Delta Hotel is in the middle.

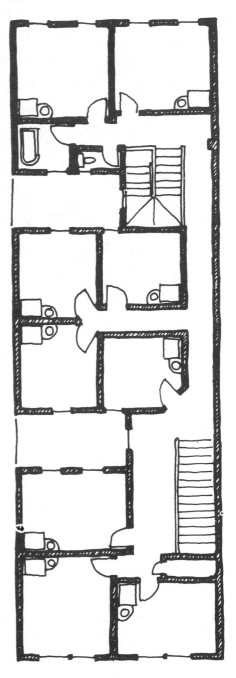

Figure 11.5 Second floor plan, Delta Hotel, San Francisco.

0   5   10

N

Figure 11.6 The Waco Hotel in St. Louis; middle-class in the 1860s and a workers' hotel when photographed in 1939. (Farm Security Administration, Library of Congress.)

However, as many as one half of all single-room tenants could not afford even an old, dilapidated rooming-house type of hotel. These people, overwhelmingly single men, were migrant workers who wintered in the city or workers who had given up migrating and settled to do casual day labor. They not only washed dishes and peeled potatoes at the better hotels, but also built the city's utility and railroad networks and helped factory owners meet peak seasonal demands.[8]

As the numbers of these people mushroomed with the industrial expansions during and after the Civil War, they usually found lodgings in temporary, left-over, or surplus space: in cast-off hotels, hastily built barracks, or partitioned cubicles constructed within generic loft spaces (Figure 11.6). Most observers called these dwellings "cheap lodging houses." As had happened with rooming houses, the pressures of codes and the demonstration of clear profitability eventually encouraged investors to provide more purpose-built lodging houses. By 1910 virtually all the new examples were permanent, not ad hoc buildings.

New or old, everything about cheap lodging houses reminded the residents that they were members of a category, not individuals. The floor plans of long halls and bare stairs did nothing to hide the facts of 100 or more tiny spaces per floor. At the lowest price level, rooms rarely had sinks and often had no closets; they simply offered hooks on the wall, a chair, and a bed. Cleanliness was a concept, not an experience. Only flophouses (where men slept on the floor) and charity missions cost less. Yet the street façades of these architect-designed structures resembled large versions of downtown rooming houses.

At *all* types of hotels—from palace hotels to flophouses—the staff worried about the inevitable intermixture of thieves, gambling sharpers, unmarried couples, and drug dealers as well as drug addicts, prostitutes, and "loose women" (although few commented on the "loose men"). Such people discredited hotels as a place to live, especially for families. In fact, small rooming houses and small houses of prostitution often occupied identical buildings in the same neighborhoods.[9]

The cheaper hostelries often sheltered chronic drunks or tenants who were marginally employable due to work injuries or mental illness. Managers could legally turn such people away, but often a hotel's dilapidated physical or financial condition warranted catering to them in order to break even or perhaps make handsome profits.[10] Thus, although marginal people hurt hotel reputations, they were a defensible, dependable, and profitable market for slumlords and honest landlords alike.

As the price-ranked hotel buildings showed, the builders and managers of residential hotels saw and counted the people who were unsatisfied by apartments, tenements, and suburban houses or who were unable to rent them because of cost or lack of social standing. Yet from the very beginning of the Progressive Era, in the 1880s, this commercial vision of hotel people differed sharply from that of public architects.

The Public Architects' Vision: SRO Residents as Pariahs

Essayists, social critics, social workers, housing reformers, and opinion leaders in the design professions saw the people who lived in hotels not as potential clients or as important to the burgeoning downtown labor supply but as cultural threats. From the Civil War to the Depression, closely interwoven groups of reformers were designing a new kind of urban culture to match the new challenges of metropolitan America. In the independent, cosmopolitan, and footloose atmosphere of hotel life, reformers perceived a strong danger to the ideals of a purer residential life that they were learning and promoting in new railroad and streetcar suburbs.

Most influential were the opinions of the New York City housing reformer Lawrence Veiller, who translated Jacob Riis's moral consciousness into systematic research and legislation. Veiller became the nation's first professional housing reformer. From the 1880s to 1917 he stood as the nation's preeminent housing expert—and thus, a public architect par excellence.[11] In Veiller's view, the residents in any kind of hotel were not worth seeing, at least not clearly. In an address to housing reformers he asserted that the "Waldorf Astorias at one end of town and the 'Big Flats' [tenements] at the other are equally bad in their destruction of civic spirit and the responsibilities of citizenship."[12]

A later public architect, Henry Wright, held similar but more temperate views. He explained the typical apartment family as inferior to the separate families in single-family houses; with a sneer he compared apartment dwellers to hotel residents:

> [The inception of the apartment house] was brought about by the vaguely defined need of families whose habits were of a semi-transient nature or those who did not care to assume the responsibilities of the independent house, and yet neither desired nor could afford the luxury of hotel life.[13]

Public architects such as Wright mistrusted middle-class families who did not buy furniture and tie themselves into place with mortgage payments. He also seemed unaware of the family life in spartan hotels.

At the cultural core of these criticisms burned the conviction that hotel life undermined Victorian norms for family togetherness and middle-class gender roles. Furthermore, proximity to commercialism, gamblers, prostitutes, or unmarried people tainted the hotel home. In 1903 the editors of *Architectural Record* railed that communal or commercial dining in hotels was the "consummate flower of domestic irresponsibility"; it meant the "sacrifice of everything implied by the word 'home.' No one," said the editors, "could apply the word to two rooms and a bath."[14]

Privacy, not the sharing of space, reigned as a molder of character according to the public architects of the Progressive Era. The private kitchen loomed large in the reformers' rules because it made possible the key ritual of family life—dining together—and because cooking (or supervising the servants' cooking) was a key role for women. The *Architectural Record* editors wrote in detail how a middle-class or upper-class woman in a hotel or apartment hotel lost her essential personal control:

> She cannot have food cooked as she likes; she has no control over her servants; she cannot train her children to live in her particular way; she cannot create that atmosphere of manners and things around her own personality, which is the chief source of her effectiveness and power.[15]

Even if they were mothers, hotel women fell short on half of true motherhood as Americans had come to define that concept during the 1800s.[16]

A separate bathroom also figured prominently in the issue of proper privacy. In his basic reference work of 1910, *Housing Reform*, Veiller insisted that each family have its own separate water closet. When toilets were located in public parts of a multiple dwelling, he said, "the use of such conveniences led to serious social evils and should not be tolerated in future buildings."[17] Significantly, Veiller called the shared toilet a *social* and not a *health* issue.

The negative official opinions about hotel people varied according to social and economic class. Public architects saw wealthy and middle-class hotel residents as second-class citizens, but as citizens nonetheless. Reformers considered the rooming house men and women as *potential* citizens, perhaps, but people in an abnormal and temporary household state of being single. Furthermore, rooming house districts often had numerous cafes, bars, and dancing establishments, which raised concerns for workers' morals and associated dangers of a "too individualistic" life.[18]

When they considered migrant and casual workers, however, public architects did not see potential citizens but *only* tramps and bums. In 1913 a California health official, after several years of on-site inspection of some of the worst hotel health offenses (Figure 11.7) gave this description for the residents of San Francisco's cheapest lodging houses:

> The inmates of these places are human wrecks, old and young, diseased or habitual drunkards; others who have failed in the struggle for existence, and those who from the beginning were never able mentally or physically to compete with their fellow men, their condition making it all the more urgent that they be compelled to live in a more sanitary environment.[19]

Although a handful of public architects wrote vigorously to balance this monolithic and negative view of migrant laborers and casual laborers, their efforts at balance failed.[20]

Small wonder that well-meaning designers and public servants found it hard to see positive attributes in these people they called "homeless men," people whose values of cleanliness, sobriety, self-control, steady employment, material possessions, and commitment to work and family not only were alien to most Americans, but also were assumed to be abhorrent.[21]

All hotel residents—including those in the aristocratic palace hotels—quickly came to be conflated with the worst lodging-house bums. In Veiller's *Model Tenement Housing Law*, published by the Russell Sage Foundation in 1910, a fateful twist in the professional view appears. Following his definition of *tenement* (which stood for both tenements and apartments, following the precedent of earlier New York City ordinances), Veiller emphasized the following point:

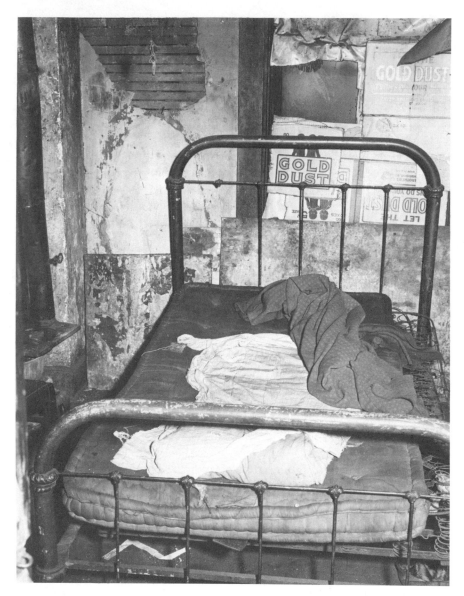

Figure 11.7 True reason for reform: a bed photographed in a residential hotel for black Americans in Chicago, 1941. (Farm Security Administration, Library of Congress.)

[The definition of tenement] does not include hotels or lodging houses. It should not do so. The problems of the *common lodging house* occupied by *homeless men or homeless women* are totally different from the problems of the tenement house occupied by families. The two should not be confused.[22]

Veiller's twisted dictum shaped and reflected the ideas of the typical public architect's vision of hotel life: any kitchenless or single-room occupants meant the worst possible case—bums in lodging houses, probably with an admixture of thieves, drug addicts, and other marginal types. Thereby, even if they had lived comfortably and by choice in the same dwelling for many years, the residents of the Waldorf Astoria or reasonable downtown rooming houses became "homeless men and homeless women." The public architects' ideology saw them as nonpeople: the antithesis of what American housing tenants were supposed to be.

The Twisted Optic Nerve: Behind the Public Architects' Rhetoric

The general notions of environmental determinism, desire for a monolithic culture, and fear of mixtures actively fed the professional attitudes against hotel life. Determinism probably figured most strongly. At the turn of the century, social scientists were convinced that bad environments directly caused psychological, moral, and social problems. The first president of the San Francisco Housing Association, a doctor, stated flatly that "health of the individual, physical and moral, and health of the community, physical and moral, both depend in no small degree on the dwelling in which the individual is housed."[23] The close correlation of bums and nearby lodging houses seemed to confirm the reformers' mistrust.

Compared with environmental determinism, the desire for a monolithic culture worked more quietly but even more widely. Americans of the Progressive Era prized individuality like that of Thomas Edison or Theodore Roosevelt, but they did not widely value pluralism, heterogeneity, or diversity as related cultural attributes. Along with the upsurge of American nativism at the turn of the century came concern about the serious cleavages between immigrant and native-born, poor and rich, single and family-tied. With campaigns to "uplift," "modernize," and "Americanize," reformers sought to transform home environments—not only to promote individual health but also to promote cultural homogeneity and usher in an ideal classless society.[24] Social workers imbued with middle-class ideals attempted to teach immigrant mothers how to decorate their homes (colonial revival or mission furniture with no doilies), how to have formal parlors, and how to avoid "wrong" foods such as spaghetti.

This parlor-and-kitchen crusade dovetailed with the City Beautiful movement, where architects, city planners, and their Chamber of Commerce backers worked for a monolithic culture at the scale of entire city streets. With new city-building rules—uniform building standards, coherent facades, and imperial beauty on wide new boulevards—public architects hoped

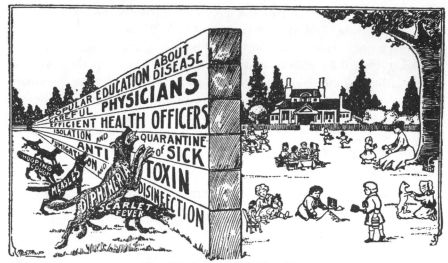

How High is the Wall in Your Town?

Figure 11.8 Health poster of 1910; for planners, the howling wolves would have been "mixed use" and "rooming houses." (From Ehrenreich and English, 1964.)

to counter the chaotic symbols of individualistic, diverse, and contentious commercial values of the old city.[25]

In these programs for a uniform social and architectural culture, hotel people stressed several nerves. Public architects forgave new immigrants for not wanting traditional WASP household life, assuming that even single immigrants would eventually yearn for American family-household ways. Most hotel dwellers, however, including those at the cheap lodging house level, were people born in America, usually of northern European stock. They had left and often consciously rejected farms and traditional houses. For this, reformers could not forgive them.[26]

The public architects' fear of mixture stemmed from sanitary notions and desires to stabilize real estate values and industrial efficiency. Specialized and separate areas for every phase and class of urban life aided the designers' visions for a planned and controlled physical culture. The worst aspect of the slums, said reformers, was the "indiscriminate herding of all kinds of people in close contact."[27] Sociologists labeled the lobbies and hallways of hotels as "social mixing bowls"; the downtown rooming houses and hotels blatantly mixed stores and residences, too.[28]

Concepts of social hygiene and moral contagion—spliced into planning practice from the science of biological hygiene—further undergirded the fear of social and land use mixture. Lawrence Veiller wrote, "We know

now . . . that poverty too is a germ disease, contagious even at times; that it thrives amid the same conditions as those under which the germs of tuberculosis flourish—in darkness, filth, and sordid surroundings."[29]

Zoning, land use planning, and tight building controls promised to mitigate the dangers of mixture, as public health posters promised (Figure 11.8). From the howling wolves of social and biological disease, civic control and careful separation would preserve the family in its protected realm of house and open yard.[30] However, public architects were convinced that hotel people—separated or not—incubated the social diseases of density, eccentricity, and mixture that would continually threaten the health of the ideal social body. Because *all* hotel residents were seen as homeless people, *all* hotel homes would undermine the brave new city to be built for the mainstream family.

Vision as Projection: Seeing Ideal People for Ideal Houses

The needy housing tenants that planners saw most clearly were family groups living in the shanties, cottages, and large tenement buildings in America's urban slums. In this as in so many other aspects of the Progressive housing movement, Lawrence Veiller set the tone and the vocabulary, with architects and planners later chiming in. When Veiller spoke of "labor" and "working men" he meant laboring men with families; he rarely mentioned working women or single mothers, and never single men.[31]

In 1902 Veiller inspected housing conditions in many large American cities. In San Francisco the poorest living quarters that he recognized were the "homes of the longshoremen, situated near the docks . . . containing two families each." Veiller most certainly walked past dozens of rooming houses and lodging houses for single longshoremen, who also lived near the docks. But Veiller did not consider them a legitimate part of the housing problem. He concluded, in fact, that San Francisco (whose Chinatown and Barbary Coast were among America's most notorious single people's slums) had *no* tenement house problem.[32]

The official vision was not entirely skewed in all places or in all decades. In a wide-ranging review of building projects and professional literature available after World War II, the sociologist Arnold M. Rose found a distinct curve of interest in the housing of single people, which was largely an issue of hotel housing. Before 1890, Americans showed almost no interest; projects and articles peaked from 1890 to 1915, and then from the 1920s through 1948 (when he wrote his study) Rose found virtual silence.[33]

The peak period from 1890 to 1915 saw the building of many municipal and benevolent lodging houses, YMCA's, room registry movements, and

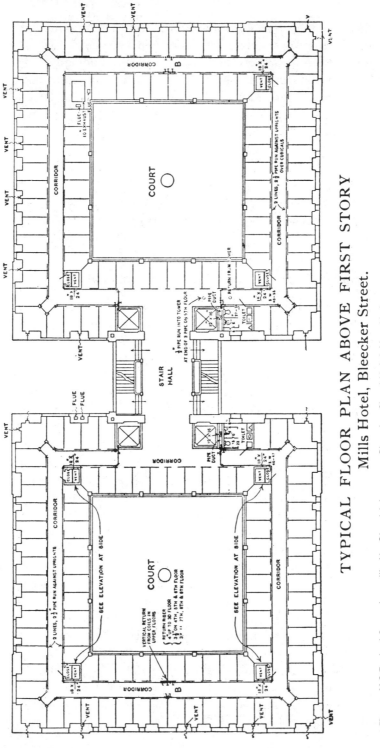

TYPICAL FLOOR PLAN ABOVE FIRST STORY

Mills Hotel, Bleecker Street.

Figure 11.9 Typical floor plan with 6′ × 9′ cubicle rooms, Bleecker Street Hotel, New York. (From Thomas, 1899.)

boarding house organizations. During this time the philanthropist Darius Mills hired the New York architect Ernest Flagg to design several model hotels for poor and single working men. Flagg's practice distinguishes him, along with Veiller, as one of the most exemplary public architects of the period. For Mill's first structure, the Bleeker Street Hotel (opened in 1897), keeping 1,500 rooms at the desired rent meant that Flagg had to accept many standards of market-rate cheap lodgings: the rooms were cubicles, with walls that did not reach the ceilings; toilet rooms were in relatively distant corners, and showers and washing areas bulked in long rows in the basement (Figure 11.9).[34]

Yet Flagg and his client agreed upon many middle-class additions: very strict supervision and behavior rules, two large lobbies whose accoutrements encouraged reading and writing, two huge atria celebrating sanitary fresh air, and dining on the premises (in a basement dining room). Flagg's design for the hotel exterior cleverly grouped the windows of each four exterior cubicles into a single composition so the large number of small and repetitive units would not be apparent on the street. The façade carefully disguised the hotel people inside behind a screen connoting an office or an apartment building (Figure 11.10).

In spite of his buildings for hotel residents, Flagg nonetheless shared the public architect's preference for family-tied people. Only four years after designing the huge Bleeker Street Hotel, he wrote that the "true interest" of state housing laws "should be to foster the type of building that best conduces to the preservation of the family and of home life, and this type can hardly be said to be the hotel."[35]

For their ideal of housing every person in a family group, reformers had a matching architectural image: as Veiller put it, "small houses, each occupied by a separate family, often with a small bit of land, with privacy for all, and with a secure sense of individuality and opportunity for real domestic life."[36] Separate bungalows with separate plots were rarely within reach of the low salaries paid to slum tenants, at least before World War II. Reformers accepted the apartment with private bath and private kitchen as a practical compromise for the minimum unit. Through the 1920s, they also had to allow new rooming houses and lodging houses, since new investments in factories, streetcars, utility systems, and office towers created thousands of jobs for hotel residents of all types.[37] Such diversity of housing provision would not last for long.

Building the Blind Spot: From Vision to Policy to Form

In spite of the last surge of residential hotel construction, starting as early as the 1870s housing professionals had been effectively inserting their anti-

Figure 11.10 Façade view, Bleecker Street Hotel, New York. (From Thomas, 1899.)

hotel blinders into official housing policies. Early health laws and then room use and land use regulations aimed directly at hotel dwellers. New York's 1910 model tenement house code, later adopted in dozens of other cities and states, classified the cheapest hotels (and by implication the residents of those buildings) as nuisances on a par with rag storage and backyard cattle.[38] The same model ordinance reiterated New York City's 1887 definition of an apartment unit: Tenement and apartment buildings had three or more families *"living independently* of one another and *doing their own cooking."* Later, writers of zoning codes used the same wording.[39]

By the 1920s, California building code writers had furthermore stipulated that kitchens were to be specialized rooms. Housing inspection forms had "cooking and sleeping in the same room" as a violation category along with topics such as "undersize rooms" and "overcrowding."[40]

As they ferreted out mixed uses of any kind, public architects also narrowed the choice of units within a single building type. For instance, early in the century a large, six-story apartment structure could have forty-five private units, each with a kitchen and private bath, and at the same time could have thirty-six one-room units (some with kitchens) whose tenants shared several baths down the hall. The developer had nestled a small bargain-priced rooming house inside a larger apartment building. Even elegant apartment blocks for the very wealthy had kitchenless units included. Model reform laws, however, discouraged this mixture.[41]

Not only making rules but also marshaling housing facts played key roles in the public architects' strategies. As Samuel Hays has observed, after 1900 the social control of self-appointed reformers depended on their monopoly of detailed surveys and professional specialization. Those who defined and quantified an urban problem gained the right to lead in the crusade against it.[42] Well-meaning reformers actively used this control of official knowledge against hotel life. Even during the professionals' early wave of interest (which crested about 1915), for every large book about families in tenements, usually only a slim article appeared about people in hotels. By the 1920s that token measure dwindled as well.[43]

The logic of ignoring both hotel residents and their homes closely resembled the logic of the Volstead Act, which legislated the national prohibition of alcoholic beverages. Reformers assumed that if they took away the *supply* of a mistrusted part of the material culture, its *use* and *need* would end as well.

Although use and need of residential hotels certainly did not end, public architects did begin to accumulate an official and deliberate ignorance about hotel residents. Because of the gap in 1920s work, the next generation of housing experts inherited a blind spot. In the New Deal, the wealth of excellent studies, reports, outlines, discussions, and conferences about housing problems included almost no information about single-room housing. At best, only the so-called homeless men of cheap lodging houses received attention.

In 1936, during some of this century's most important meetings of public architects about housing issues, reformers codified the *Property Standards* for the new Federal Housing Administration. Predictably, hotel residents moved even further out of the official cone of vision. According to the FHA, a legal housing unit *had* to include a private kitchen and private bath. The guidelines made a more general category of *dwelling*—"any structure used principally for residential purposes"—but carefully stated that commercial rooming houses, clubs, or other single room housing would not be considered even as dwellings within the meaning of the National Housing Act that had created the FHA.[44] These policies, based on forty years of public architects' skewed vision of SRO people, blocked federal attention and funding for hotels for forty-five years. The U.S. Bureau of the Census, which in 1930 published its first attempt at a total hotel room count, had never counted the subset of residential hotel units, and to this date never has.[45]

The FHA guidelines almost immediately affected the New Deal's Real Property Surveys, which set the geographic agendas for later urban renewal projects. In San Francisco's survey of 1939, a small army of field enumerators collected excruciatingly detailed room-by-room interviews from hotel tenants as well as from occupants of apartments, flats, and houses. However, erasures and editing on the original forms show that the office staff systematically excised hotel people from their compilations. Buried in the appendixes of their report they hid the fact that one third of the city's substandard dwelling units and ill-housed people were in hotels.[46]

When data gathering gave way to actual building, the omission of people changed to omission of their homes. Soon after completing the Real Property Survey, San Francisco's city planners tagged residential hotel blocks as sites for freeway construction and other massive land use changes. When it came time to clear a block, laws required the government to help relocate people in apartments; they were real people in real housing. But if the project displaced hotel dwellers they were neither counted nor relocated; they were nonpeople in nonhousing.

In the 1950s, just as in the Progressive Era, a small minority of public architects had not developed the blind spot about SRO people. In 1953, for instance, Catherine Bauer and Davis McEntire recommended that the Redevelopment Agency of Sacramento, California, construct a *new* single workers' district closely modeled on the old West End skid row. The notion of publicly built and privately managed rooming houses above stores met far too many obstacles for the startling recommendation to be taken seriously.[47]

The public architects' blind spot continued to grow. In 1972, the *Journal of Housing* reported that FHA and HUD would finally consider shared kitchens but not shared bathrooms in what they called "congregate housing." In 1976 and after, HUD officials continued to refuse funding of hotel housing.[48]

Meanwhile, from the 1950s to the mid-1970s, in the most massive building burst in American history, architects and planners tore down hundreds of thousands of residential hotel rooms along with other, counted housing units. Since 1960, Seattle has lost 16,000 downtown housing units; Portland, Oregon, over 2,400—most of them SROs. New York City had over 50,000 lower-priced SRO units in 1975, and only 19,000 remained in 1981, a decline of 61 percent.[49]

Overall, through the withholding of funds to repair old residential hotels, refusing to build new SRO units, and tearing down both good and bad existing units, public architects have built their blind spot about SRO people into a physical vacuum in downtown America. Congress finally changed the relocation policy in 1970, but not until the 1980s have a few federal dollars trickled into a handful of single-room housing projects. Officials still consider the projects highly experimental.

Public Architecture and Social Vision Impairment

The building of the SRO blind spot points to the power of ideology and social vision in the planning and design process. As Peter Marcuse and Mary Burki have shown in recent analyses, a cultural imperialism in the vision of people has haunted the public housing movement from its inception. At the office, planners repressed or ignored their own cultural biases and how these biases skewed their definitions. On the street, planners ignored or remained unaware of the different subcultures of hotel housing tenants. Hence, the planners and architects defined the ideal residential environments of both the suburbs and downtown based largely on their own experiences and professional design perspectives.[50]

Assuming that their vision of the world is complete and accurate is of course an essential factor in the practice of public architects. The cultural imperialism pointed out by Marcuse and Burki manifested itself in the development of *spotlight vision*—an impairment of social view that helped public architects act like the utopian designers they wanted to be rather than the social servants they rationalized themselves as being. When called upon to define their clients, they did not take in the reality of the city's social and cultural diversity. Instead, from inside their own minds, reformers projected out the bright, shining image of the ideal urban household, the "ought-to-be" family. They proceeded to project the single-family house, privatized apartment, and single-use district as the only solutions to metropolitan residences. They then wrote these views of people into the national definition of housing units.

Spotlight vision also projected the definition of the "homeless" person. Even if people had lived happily and productively in a hotel for fifteen years, if they did not have a private kitchen and bath, they were "homeless." The image came from within the planners as much as from true vagrants or street people.

The historical conflict between reformers and SRO life elucidates other major types of social vision impairment that can strike public architects. *Filtering*, like the wearing of dark sunglasses, admitted information about some sectors of the urban population—immigrants and particularly those living in family groups—but not others. Often the filter itself matched the pattern of burns left by spotlight vision. For instance, the narrow definitions of housing units curtailed the observation and counting of SRO tenants.

Soft focus, the visual equivalent of overgeneralizing in speech, blurred the view of one social group into another, to make them all a single lump. Public architects saw all single men living in cheap rooms near skid row as drunks and bums, and all moneyed hotel people as an irresponsible "gilded, gabby gang of newly gotten rich."[51] Such blurriness reduced the possible encouragement of hotel life.

Double vision occurred in public architecture when designers had to work against their own ideals. Although Lawrence Veiller rarely had to contradict his own views in his work, architects were often caught on both sides of the hotel housing issue. Ernest Flagg's own theoretical schizophrenia points to the personal problems: He designed thousands of single-room dwellings for poor workers while elsewhere completely disparaging hotels as homes.

Blind spots developed where spotlight vision had torn a hole in the mental retina, where filtering had deprived information to a section of the brain, or where soft focus had not been corrected over a long period of time. The blind spot about SRO people began with seeing people negatively or in blurred clumps, then leaving them out of reports, and finally ignoring them because they had ceased to exist in the official records.

With these dysfunctions of social vision—in reality, disconnections between eye and brain—the lawyers and social workers and architects of the early housing movement managed to make hotel residents into nonpeople. The ideological problems of public architects versus hotel residents reveal the inextricable relation between vision and action in public design.

The visual and conceptual dysfunctions that lurk in the practice of public architecture arise from the struggles with messy social heterogeneities that poke holes in elegant but simplistic design schemes. In building or managing public space, one bumps into two central realities: first, people are diverse; and second, diversity requires multiple solutions and flexibility. As difficult as a multiple social vision is to acquire, it is essential for public architectural practice.

The human costs of the public architects' ideology about hotel people have been enormous. They have set into place formidable amounts of public apathy and private inertia. As a new group of activists has begun to fight for SROs and to push them into the public architect's cone of vision, they continue to encounter the mass of textbooks, laws, bureaucracies, and attendant attitudes that retard changing the definitions of "public housing client" and "dwelling unit." The vision *is* widening, but slowly.[52] In the future, the more that professional vision can be wider, more inclusive, and more pluralistic, the less likely that we will repeat the human costs of the American SRO example.

Notes

A grant from the Beatrix Farrand Fund (Department of Landscape Architecture, University of California at Berkeley) and the assistance of Louise Mozingo and Leslie Watson significantly aided the preparation of this chapter.
  1. California state law sets a six-room limit; other states set from four to ten rooms as the limit.
  2. The general review in Jefferson Williamson, *The American Hotel: An Anecdotal History* (New York: Alfred A. Knopf, 1930) is extended by Norman Sylvester Hayner, "The Hotel: The Sociology of Hotel Life," Ph.D. dissertation in sociology, University of Chicago, 1923.
  3. The newspaper illustration (Figure 11.2) is in Leslie Dorsey and Janice Devine, *Fare Thee Well* (New York: Crown, 1964), p. 139.
  4. Day Monroe, *Chicago Families: A Study of Unpublished Census Data*, Social Science Studies, no. 22 (Chicago: University of Chicago Press, 1932), esp. pp. 66–67; "New First Lady Is 'Human,' Has Sense of Humor," *Chicago Daily Tribune* (August 4, 1923).
  5. See Lucius M. Boomer, *Hotel Management: Principles and Practice*, 2d ed. (New York: Harper & Brothers, 1931), p. xvii.
  6. The best available study on rooming houses is Albert Benedict Wolfe, *The Lodging House Problem in Boston* (Boston: Houghton Mifflin, 1906); on these issues, see pp. 6, 38, 42–44, 94. See also John Modell and Tamara K. Hareven, "Urbanization and the Malleable Household: An Examination of Boarding and Lodging in American Families," *Journal of Marriage and the Family* 35 (1975):467–79.
  7. Paul Groth, " 'Marketplace' Vernacular Design: The Case of Downtown Rooming Houses," in *Perspectives in Vernacular Architecture*, ed. Camille Wells, vol. II (Columbia: University of Missouri Press, 1986), pp. 179–91.
  8. For a positive contemporary view of migrant laborers, see Nels Anderson, *The Hobo: The Sociology of the Homeless Man* (Chicago: The University of Chicago Press, 1923).
  9. For a sampling of the literature on these groups, see Paul Groth, "Forbidden Housing: The Evolution and Exclusion of Hotels, Boarding Houses, Rooming Houses, and Lodging Houses in American Cities, 1880–1930," Ph.D. dissertation in geography, University of California, Berkeley, 1983, pp. 191–203, 340–427.

10. For a poignant and all too common description of a rooming-house keeper's dilemmas, see Wolfe, *Lodging House Problem*, pp. 67–71.
11. Roy Lubove, *The Progressives and the Slums: Tenement House Reform in New York City, 1890–1917* (Pittsburgh: University of Pittsburgh Press, 1962), pp. 117–18, 128–29, 140, 148.
12. Lawrence Veiller, "The Housing Problem in American Cities," *The Annals of the American Academy of Political and Social Science* 25 (1905):248 ff.; quotation on pp. 255–56.
13. Henry Wright, *Rehousing Urban America* (New York: Columbia University Press, 1935), p. 63.
14. "Over the Drafting Board, Opinions Official and Unofficial," *Architectural Record* 13 (January 1903): 89–91. Technically, the editors were decrying the apartment hotel in this editorial.
15. Ibid.
16. For an early statement, see Barbara Welter, "The Cult of True Womanhood: 1800–1860," *American Quarterly* 18 (Summer 1966):151–74.
17. Lawrence Veiller, *Housing Reform: A Handbook for Practical Use in American Cities* (New York: Charities Publication Committee, 1910), p. 109.
18. Arnold M. Rose, "Living Arrangements of Unattached Persons," *The American Sociological Review* 12 (1947):429–35; on hotel-district dangers, see Wolfe, *Lodging House Problem*, p. 154, and Harvey Warren Zorbaugh, *Gold Coast and Slum: A Sociological Study of Chicago's Near North Side* (Chicago: University of Chicago Press, 1929).
19. Mrs. Johanna von Wagner, of the California State Tuberculosis Commission, quoted in San Francisco Housing Association, *Second Annual Report* (San Francisco: The Association, 1913), p. 24.
20. In addition to Anderson, *The Hobo*, see William T. Stead, *If Christ Should Come to Chicago* (Chicago: Laird & Less, 1894), p. 30. See also the 1950s work of Catherine Bauer, cited later.
21. Spradley, *You Owe Yourself a Drunk: An Ethnography of Urban Nomads* (Boston: Little, Brown and Co., 1970), pp. 6–7, 67; Samuel E. Wallace, *Skid Row as a Way of Life* (Totowa, N.J.: Bedminster Press, 1965), pp. 129, 144.
22. Lawrence Veiller, *A Model Tenement House Law* (New York: Russell Sage Foundation, 1910), pp. 14–15 (emphasis added).
23. Langley Porter, M.D., quoted in San Francisco Housing Association, *First Report* (1911), p. 6, 8.
24. On nativism, start with John Higham, *Strangers in the Land: Patterns of American Nativism, 1860–1975* (New Brunswick, N.J.: Rutgers University Press, 1955); on house interiors, see Lizabeth A. Cohen, "Embellishing a Life of Labor: An Interpretation of the Material Culture of American Working-Class Homes, 1885–1915," *Journal of American Culture* 3 (1980):752–75; my phrasing draws directly from Cohen, pp. 756, 761.
25. A well-focused study examining these ideas is Judd Kahn, *Imperial San Francisco: Politics and Planning in an American City, 1887–1906* (Lincoln: University of Nebraska Press, 1979). See especially pp. 3, 215–16.
26. An apt and acerbic example can be found in Edith Abbott, *The Tenements of Chicago, 1908–1935* (Chicago; University of Chicago Press, 1936), pp. 343, 345–46.
27. Robert W. Deforest and Lawrence Veiller, eds., *The Tenement House Problem* (New York: Macmillan, 1903), vol. 1, p. 10. The writer is Veiller.
28. Norman Sylvester Hayner, *Hotel Life* (Chapel Hill: The University of North Carolina Press, 1936), p. 93.
29. Veiller, *Housing Reform*, p. 5.
30. The health poster view of these ideas is in Barbara Ehrenreich and Deirdre English, *Complaints and Disorders: The Social Politics of Sickness* (Old Westbury, NY: Feminist Press, 1973), p. 64.

31. See, for instance, Veiller in Deforest and Veiller, *Tenement House Problem*, Vol. 1, p. 3; and Veiller, *Housing Reform*, p. 155.

32. Deforest and Veiller, *Tenement House Problem*, vol. 1, p. 144.

33. Arnold Rose, "Interest in the Living Arrangements of the Urban Unattached," *American Journal of Sociology* 53 (1948):483–93.

34. On Flagg's designs for the Mills Hotel No. 1 (first named the Bleecker Street Hotel and later renamed the Greenwich Hotel), see John Thomas Lloyd, "Workingmen's Hotels," *Municipal Affairs* 3 (1899):73–94.

35. Ernest Flagg, "The Planning of Apartment Houses and Tenements," *Architectural Review* 5 (New Series; vol. 10 in the old series; August 1903):85–90; quotation is on p. 89.

36. Veiller, *Housing Reform*, p. 6. See also Groth, "Forbidden Housing," pp. 535–40.

37. In fact, during World War I the federal government's first public housing projects (for the U.S. Shipping Board's Emergency Fleet Corporation) included dormitory buildings and cafeterias for single workers.

38. Veiller, *Model Tenement House Law*, Section 94 (p. 80); this language, with no changes, became Section 70 of California's first state tenement house law (California Statutes of 1915, Chapter 572).

39. On New York, see Lubove, *Progressives and the Slums*, pp. 18, 26; and Veiller, *Model Tenement House Law*, Section 2, pp. 13–14, (emphasis added). San Francisco's 1921 and 1927 zoning ordinances followed suit; see San Francisco, City and County, *Building Zone Ordinance and Zone Maps* (Ordinance No. 5464, with amendments to January 1, 1927), Section 1.

40. California Statutes of 1917, Chapter 736 (Hotel Act), Section 65. Forms with the noted categories are in the 1920s microfilm files of the San Francisco Department of Apartment and Hotel Inspection.

41. This example stands at 172–180 Sixth Street in San Francisco. For an upper-class example, see the plans for 277 Park Avenue, New York, designed by McKim, Mead, and White in 1925, on p. 91 of Robert A. M. Stern, "With Rhetoric: The New York Apartment House," *Via 4: Culture and the Social Vision* (Philadelphia: University of Pennsylvania School of Design, 1980), pp. 78–111. On a model law written in 1929, see James Ford, *Slums and Housing: History, Conditions, Policy with Special Reference to New York City* (Cambridge: Harvard University Press, 1936), vol. 1, p. 348.

42. Hays in the foreward to Lubove, *Progressives and the Slums*, pp. x–xi.

43. See Rose, "Interest in Living Arrangements."

44. Federal Housing Administration, *Property Standards: Requirements for Mortgage Insurance under Title II of the National Housing Act, June 1, 1936* (Washington, D.C.: Government Printing Office, 1936), pp. 14–15, 17.

45. Hayner, "Sociology of Hotel Life," pp. 69–70.

46. The buried proportion is on Tables C and D under "Project Operations," in San Francisco, Housing Authority of the City and County, *Real Property Survey, 1939* (San Francisco: Housing Authority, 1940), vol. 1, pp. 288–89.

47. Catherine Bauer and Davis McEntire, "Relocation Study, Single Male Population, Sacramento's West End" (Sacramento: Redevelopment Agency of the City of Sacramento, Report 5, 1953), pp. 2–5.

48. See Dorothy Gazzolo, ed., "Skid Row Gives Renewalists Rough, Tough Relocation Problems," *Journal of Housing* 18 (August–September 1961):327–36, and Byron Fielding, "Low Income, Single-Person Housing," *Journal of Housing* 29 (April 1972):133–36. For a typical HUD official's statement, see B. Joyce Stephens, *Loners, Losers, and Lovers: Elderly Tenants in a Slum Hotel* (Seattle: University of Washington Press, 1976), p. 25.

49. On Seattle, U.S. Senate Special Committee on Aging, *Single Room Occupancy: A Need for National Concern* (Information Paper, June 1978), p. 4; on Portland, Samuel Galbreath speaking at "Residential Hotels: A Vanishing Housing Resource," a conference

sponsored by the Governor's Office of Planning and Research and the California Department of Housing and Community Development, held in San Francisco on June 11–12, 1981; on New York City, Susan Baldwin, "Salvaging SRO Housing," *City Limits: The News Magazine of New York City Housing and Neighborhoods* (April 1981):12–15.

50. Peter Marcuse, "Housing in Early City Planning," *Journal of Urban History* 6 (February 1980):153–77; Mary Ann Burki, "Housing the Low-Income Urban Elderly: A Role for the Single Room Occupancy Hotel," Ph.D. dissertation in urban and public affairs, Portland State University, 1982, p. 286.

51. Elbert Hubbard [bungalow-living promoter and editor of *Craftsman* magazine], "A Little Journey to Hotel Sherman," n.d., quoted in Hayner, "Sociology of Hotel Life," p. 159.

52. The conference held by the California governor's staff, cited previously, was a major demonstration of changing concerns and experimental housing projects in many cities; for other samples, see J. Kevin Eckert, *The Unseen Elderly: A Study of Marginally Subsistent Hotel Dwellers* (San Diego: The Campanile Press, 1980); John K. C. Liu, "San Francisco Chinatown Residential Hotels" (San Francisco: Chinatown Neighborhood Improvement Resource Center, 1980); Bradford Paul, "Rehabilitating Residential Hotels," Information Sheet No. 31 (Washington, D.C.: The National Trust for Historic Preservation, 1981), 46 pp.; Frances E. Werner and David B. Bryson, "A Guide to the Preservation and Maintenance of Single Room Occupancy (SRO) Housing" [The SRO Housing Guide], *The Clearinghouse Review*, published in two parts: Part 1 (April 1982), pp. 999–1009, and Part 2 (May 1982), p. 1–25.

12

Postmodern Architects' People

GIANDOMENICO AMENDOLA

Postmodern architecture is a residual concept defined by a principle of contrast. Like most definitions of "post" phenomena (e.g., "postindustrial"), it is characterized by its indefinite and exploratory nature. Such a form acquires meaning through comparison to a preceding moment and by marking changes indicative of the surpassing of that moment. It furthermore can only be understood by examining the underlying assumptions about the people and society in whose name it claims legitimacy.

In spite of its great array of styles and forms, postmodern architecture constitutes a particular movement, as well as an analytical category, precisely because its many representatives and versions have shared an underlying need to find design solutions to the problems that made the rationalistic movement fail. Born out of the stylistic and social crises of the modern movement, postmodern architecture took shape as the CIAM architects responded to the inadequacies of their own erroneous cultural and sociological assumptions. But the lack of meaningful communication, the users' practical and symbolic dissatisfaction, and the uncertainty of clients are due as much to the profound social changes that have taken place in the fifty years since the birth of the modern movement as they are to these basic erroneous assumptions.

The sociological assumptions underlying the success and the failure of the modern movement were grounded in a rationalistic model of progress with three corollaries: (1) a relentless tendency to homogenize values and communication codes; (2) unilinear modernization and cultural homogenization organized around universal values of rationality with respect to the ends (Weberian *Zweckrationalität*) that could be evaluated in scientific criteria; (3) the elimination of differences on an international scale, in accordance with functionality and with scientific criteria.

The respect for "truth" and "science" guaranteed the ability of architecture to communicate and transmit reduced complexity.[1]

By appealing to these principles, Le Corbusier defended his own Western and universalistic design and planning criteria from local critics who stressed the colonial approach of his project for Chandigarh in India. His dwelling units in Marseille, created according to scientific rationality (they were machines), required that people adapt through a learning process. Pedagogy was the keystone of the whole system, and an efficient socialization process was the prerequisite of design effectiveness. The matching of the users' needs to design was assured, on the one hand, by the scientific approach and, on the other, by the rational bias of the user. The user was assumed to be a rational-Smithian-*Homo oeconomicus*, free of atrophied "residues" or local and particular cultural ties.

The universality of "scientific" design and the "rationality" of users denied the problem of the users' differing social strata: architectural design presumed and served a universalistic society made up of members of equal class, status, culture, and knowledge.

Finally, the modern movement assumed—as the illuministic tradition had—that the antithesis between the individual and society could be easily overcome. It was believed that individual reason could be brought to conform with universal and societal reason. Architecture, to allude to Le Corbusier again, would avoid revolution.

The circumstances regarding the failure of the modern movement in architecture have been broadly discussed. They include loss of meaning of the built environment; inability of the architects to individuate and respond to the needs of the users; a crisis in the ability of architects to communicate, stemming from their ignorance or unwillingness to acknowledge the existence of linguistic codes, cultural models, and levels of competence. The simultaneous result of users' dissatisfaction and the inability of the buildings to achieve their purpose is clearly evident in the 1950s and 1960s, a time of sweeping changes in the "affluent" society and the so-called revolution of rising expectations. In time, the heirs of rationalism became self-conscious of the social nature of their failure. Responding to the error of their sociological assumptions, they began to thematize—in a still problematic way—social issues. The first, and most important, step was to assume people as a problem. For it is people, with their own real needs, who cast doubt on the laboratory functionalism of the modern movement. The variety and diversity of cultural models, identity and intentions, aspirations and competences, social trajectories and linguistic codes upset the project of a unified and unifying architecture. It is the complex and segmented nature of modern and postmodern society that makes modern architecture fail. The rationalistic movement was not able to cope with real contemporary people.

Communicative Architecture

Not only is *postmodern* a label that unifies differences, but it acts as an umbrella for a variety of styles and approaches that have their origins in the attempt to respond to the crisis of rationalism. Monumentalism, neobaroque, eclecticism, pop style, and revivalism mark different moments of this response. They all come from the change in sociological assumptions in design and from a new attitude toward people and society by architects.

The stylistic currents and revivals coming from tradition, history, mass media, or main street are not at all identical; yet when interwoven, they tend to reinforce each other and constitute a postmodern approach as a whole.

The neobaroque component of postmodern architecture evokes the spectacular dimension of the built environment and in its monumentality addresses the problematic but crucial relationship between the individual and society. Neoeclecticism's aggressive mixing of styles and stereotypes (dominated by traditional and vernacular values in revival) constitutes one possible response to the need for communication and consensus in situations where the social and cultural identity of the user is uncertain. This aggressiveness and stylistic abundance of postmodern architecture is necessary to recall the attention of simmering urban man, rendered blase and indifferent by a sensory overload of stimuli. It is a symptomatic response to the continuous flow of new design models and signs produced by mass media culture and fixed in that great iconic bazaar known as the contemporary American city. The contextualism of the new architecture therefore cannot be only stylistic, it must also, above all, be cultural.

Postmodern architects seek to introduce emerging and differentiating elements while respecting the problem of insertion within the existing architectural context comprised of people with differentiated cultural models and codes, needs, behavior, and expectations.

The differences, occasionally macroscopic, notwithstanding, the many currents within postmodern architecture are balanced and unified, not so much by their common need to communicate (architecture has always had this purpose), but by their shared awareness that communication itself constitutes a problem of increasing difficulty.

The doubt about intention (to use a category *à la* Luhman) underlies a real crisis of the modern movement in architecture. This is the actor's awareness of the possibility that the effects of communicative action (in this case of the design) might not correspond to the intentions behind the same action.[2]

For those who might desire to fix a precise date for the birth of post-modern architecture, we might set it with Charles Jencks on July 15, 1972, at 3:32 P.M., the moment when Pruitt-Igoe housing in Saint Louis was knocked down by dynamite.[3] At the root of its pathogenic environment was the difficulty of communication between the design and its inhabitants: The signs embodied in the building that was meant to communicate intentions and make them effectual in orienting people's behavior remained signs without sense, signals without meaning. (For a contrasting view, see Chapter 13, pp. 260–261.)

In a world in which tastes and codes are neither unified (above all between the poles of the designer and of the user) nor taken for granted in advance, postmodern architecture defines itself as "the architecture of communication," to use an expression of Paolo Portoghesi. A praxis of design that has the masses as its receiver seeks to define itself in taking on the masses as a problem. Modern "mass" society is, paradoxically, the opposite of uniform; instead, it is made up of people who are neither unifiable nor homologous to the designer. The public can no longer be taken for granted.

Postmodern architecture tends to reconstruct on a level of lived experience *(Erlebnis)* a social and cultural unity that no longer exists and that cannot possibly be assumed. This effort to recover daily, meaningful experiences is, according to Habermas, a mandate to decolonize the destroyed *Lebenswelt.*[4] Today more than ever architecture is expected both to mediate its own codes and identity as a creative discipline and to satisfy the codes and identity of the social world of its users.

The project for merging aesthetic desire and functionality is both traditional and modern. However, it must be built from the recognition of the problematic variety of aesthetic desires and practical needs. This is the social root of the ambiguity and the contradictions of modern (or, better, postmodern) architecture of which Venturi speaks when he denies the existence of a "universal visual order," given that there is no longer a corresponding social or cultural unity by which such a visual order could be produced. "[T]he design no longer possesses an autonomous idiom, but is forced to submit creatively to the needs that reality produces."[5] Thus, post-modern architects and theorists like Venturi prefer "hybrid rather than 'pure,' compromising rather than 'clean,' distorted rather than 'straightforward,' ambiguous rather than 'articulated,' perverse as well as impersonal, boring as well as 'interesting,' conventional rather than 'designed,' . . ."[6]

The Plurality of Codes

One of the new ideas of eclectic postmodern form is that the plurality of existing codes and personal styles of observers–users and participants produces a corresponding plurality of possible foci of communication. New design must also keep in mind the differing norm and levels of education and the relations between social groups of users with a high level of Culture, traditionally defined as *Culture* with a capital *C*, or a "legitimate" culture. Architects attempting to develop suggestions from the aesthetic and preaesthetic sense of the everyday world, of the *Lebenswelt* enriched by signs of mass communication, have access to a practice, an infinite catalog of "floating signs," to quote Barthes.[7] As Picasso once remarked: *"Je ne cherche pas, je trouve."* (I don't search, I find.)

Monumentality

Scenography and monumentality were the two fundamentals of baroque architecture. Likewise in postmodern architecture (some of whose expressions are considered postbaroque) there is a constant effort to create events within urban environments that are diversified and visible. This kind of monumentality both attracts and marks built and social differences.

The effect is sometimes obtained through rarefying differences in volume, height, styles, or materials. The great contemporary effort is therefore for architects to create events within an urban and cultural environment that acknowledge the dominant contemporary architectural and civic culture, while integrating their buildings with their own symbolic roots. The new architecture must find—to paraphrase what Giedion says of Borromini's great baroque—"a new power to mold taste (space) . . . and to produce an astonishing and unified whole."[8] The building–monument, independently from the meaningful varieties of functions (from a church to a residential building, from office towers to hotels), must represent a marker within the city pattern and the urban skyline.

The effort to make a building emerge from the urban landscape and increase its visibility does not respond only to a formal need to attract the attention of the consumer. This physical distinction must serve the substantive demands of function, or of the social actor embedded in the building.

Monumentality recalls the antithetical support of the individual-to-society that characterized the baroque age. The baroque "is the great spreader of that ethic principle which compares, for the first time in history, the individual with his aspirations, his weakness, his passions to the society with its new attributes of undiscussed entity and sovereignty."[9]

Even though political democracy has generally taken the place of the absolute state of the 1600s and the citizen has taken the place of the subject, the antithetical relationship between individual and society remains. The enlightenment and utopian precepts of the architects of the modern and rationalistic movements were predicated on a belief in the possibility of overcoming this contradiction by widening individual Reason until it finally overlapped with societal and universal Reason.

The continuity between the baroque and postmodern neobaroque is found in the monumental, architectonic, and social intentions of particular stylistic forms that represent a means of emphasis and communication.

The theatrical dimension is also present, as will be noted further on, with its illusory, tricky, and imaginary elements.

The Ladder of Tastes

A postmodern architect's people are subject to social stratification, cultural models and levels of competence, the affluent nature of society, and problems deriving from the rising needs of consensus. The current doubt about the effects of communication stems from the inherent existing contradiction in an architecture that purports in democratic fashion to be for everyone but that in fact must cope with a society made up of varying levels and classes.

In the baroque age, before the French Revolution and the rise of bourgeois democracy, the existence of social classes and ranks was unquestioned. Architects simply assumed ethical and political principles that postulated "natural" social differences. They faced no conflict when they identified a narrow and selected social target for which to produce built space.

Leon Battista Alberti explicitly speaks about architectural design being tied to social hierarchy: "Certain architectural products . . . are then for the benefits of all citizens, others for citizens of greater quality and others for the citizens of the meaner sort."

In the court society of Louis XIV, forms, decors, and names of houses themselves were strictly codified, as is clearly explained in the *Encyclopédie*. Only the king and the members of the royal family could erect a great building with specific characteristics and call it a *palais*. At a lower level, a reduced size was available with a different name: *hôtel*. Lower yet, for the wealthy, middle class, was the *maison*, and below this, for the lower middle class, was the *maison particulaire*.[10]

The *ancien régime* of baroque society represents the culmination of a stratified architecture. It is an architecture that solves the problem of communication, assuming the existence of two *niveaux epistémologiques*,[11] that is, two different levels of communication based on as many linguistic codes and degrees of competence of different segments of the public.

In the monumental baroque of the *ancien régime* the two epistemological levels correspond to the principal class differences of that society: the rulers and the ruled. The first was entitled to the direct practical and symbolic use of the building. It establishes a "legitimate" symbolic use of the building, being able to gather specific meaning on the basis of referents and comparisons within the universe of the stylistic possibilities; the second, on the other hand, has no role but to perceive the exceptional nature of that building and of its inhabitants. The codes of communication and the system of signs must guarantee, in this context, awe as well as admiration.

> In the strictly hierarchical society of the sixteenth and seventeenth centuries, the contrast between the vulgar and noble becomes one of the major preoccupations of critics. Within the current opinion that certain forms and modes were really vulgar because they are liked by the lower class while others intrinsically noble because only those with an evolved taste were able to appreciate them.[12]

People were meant to look at, admire, and make up their minds across a structural social gap they were not able to bridge. People were considered either to be spectators in a great scenographic experience, such as in church, or as necessary compositional elements to complete and give life to the baroque theatrical city.[13]

The difference between the neobaroque and the baroque monument, despite their similar intentions of exceptionality and diversity, is expressed in the ways the monument presents itself to the public. The basis of this different orientation lies in the assumption of people as a problem, in the knowledge of the new social demand of built space, and, above all, in the new characteristics of affluent Western society.

The problem of the new monumentality must be considered with the new characteristics of social stratification. The need for spatial and social distinctness and the effort to convey the inhabitants' or owners' social identity remain constant. The new character of postmodern monumentality, however, stems from its attempt simultaneously to emphasize exceptionality and eliminate the irreducible demarcation of rulers from the ruled that was at the basis of the monumentality of the *ancien régime*.

The monument, to the extent it expresses an asymmetric relationship between the two subjects, must be based on communication and consensus. That is, it must be meaningful for all of the public, not only in its exceptional nature, but in its ability to communicate the specific purpose of builders through the abstract language of design. As far as architectural forms are concerned, they must be, furthermore, considered the best choice out of the range of possible stylistic options. It is the popular desire to understand architecture that generates the success of art and architectural critics who boost their popularizing function by taking the self-gratifying role of unmasking architectural codes.

The Urban Chronologies

Assumptions about social actors—people—with their different behaviors and intentions, identities and strategies, values and expectations are important in orienting architectural design in an implicit and often unconscious way.

The eclectic and ambiguous monumentality of neobaroque–post modern architecture is quite different in this respect from the coherence of the seventeenth century; it is driven by the architects' awareness of both the class nature of society and its affluent and consumerist character with its intense mobility and revolution of rising expectations.

Postmodern design has had to come to terms with growing cultural differences and the break that occurred in the so-called concentricity of life, when the existence of a common set of values and sense patterns acted as the meaningful axis of all different life experiences. Contemporary architectural design, therefore, presupposes the existence of realms of experience and meaning that are not only different between individuals but also not concentric within the same individual.

Furthermore, architectural design must cope with different chronologies of social and cultural change that are typical of an affluent society that eagerly consumes goods and values. Chronologies—to use an analytical category of the French historian Braudel—of cultural changes are slow, and those of systems of meanings are even slower. Baudelaire, the urban poet who witnessed the great urbanistic transformation of Paris during the Second Empire, said: "*Le forme d'une ville change plus vite, hélas, que le coeur d'un mortal!*" ("The form of a city changes faster, alas, than the heart of a mortal!"), so affirming that built space has a chronology that is faster paced than that of the schemes that give sense and meaning to one's experience.

The chronology of taste operates much more quickly, pushed forward by two concurrent elements: the logic of production for the leisure consumption of affluent society and the social processes that shape tastes.

Major difficulties for architects stem from this very fact: The design must reconcile the slowness of change in meaning patterns with the rapidity of change in tastes. Taste is a crucial concept, insofar as it is through taste that architecture confronts social classes and their differences. It is through the clear but thick filter of taste that architects can understand the differing distribution of social, economic, and cultural capital, the status cleavages and the group conflicts that are at the very basis of the unequal distribution of resources.[14]

Taste is an expression and instrument of distinction and discrimination among social strata. Every real or potential movement of a social group toward higher levels is expressed through behavior that tends to gain control of highly valued cultural goods. The process of gaining a new social status demands symbolic means of enforcing and demonstrating the new identity. These trends toward upward social mobility lead to a continuous antagonistic carousel of actions and reactions. The groups that feel threatened defend their own relative privileges, putting into effect new behavior and new strategies of distinction through the transformation of taste. The more intense the social mobility, the faster the chronology of taste.

Postmodern architecture's response to social dynamics deals less with the shift from ascriptive to achieved status than with the direction and speed of social mobility. Design must solicit consensus in highly mobile situations and act as a status symbol in the symbolic arena of tastes, catering to the users' rising expectations. One designs in order to fulfill the expectations and desires more than factual needs; from this point of view people's income bracket is more relevant to the architect than their social condition.

Images and Status

This status-oriented design takes on particular meaning for the buildings that fulfill conspicuous functions; their success lies in their ability to accomplish the tricky task of supplying both status and services. They must attract the largest number of users and at the same time convey an effect of distinction and privilege.

Hotels and restaurants, as well as apartment houses and condominiums, are designed by articulating the social mobility that produces rising aspirations in people, aspirations usually greater than the status actually achieved. These environments must, according to the discriminating logic underlying "legitimate" culture, offer "distinctive" opportunities for consumption and at the same time make these privileges accessible even to those who are not familiar with such opportunities.

In this way the objective of contemporary architecture is to allow all the public—even that part not very familiar with legitimate culture—to be able to make use of and enjoy these conspicuous built environments. The constant attempt by the architect is therefore to organize a system of signs that in a hidden way makes users "able to recognize the signs which are worthy of admiration."[15] Making people believe—through their practice—that their capabilities match their aspirations, that they can be what they long to be, is the objective of postmodern architects.

> Most of the people coming to the hotel were looking for something flamboyant. Most of the people coming to the hotel were not especially literate. These weren't essentially cultured people. Some of them may have not been, but they had forgotten their culture. They weren't coming to Miami Beach for culture, they were coming for a vacation.[16]

The formal language of architectural design tries to prevent the user from feeling foreign and uneasy in relation to the building and encourages him to enjoy in full this spatial and social experience. He should feel at home and experience himself as the privileged connoisseur of an exclusive system of distinctive signs. Buildings should be not only functionally but most of all symbolically accessible, allowing the user to take possession of the functions and codes symbolically embedded in the space, to choose proper behavior and to acquire the social attributes encoded into the building's form.

Tricks and Culture

One of the principal instruments for conferring a sense of status is the "culture effect." The "distinction strategy" is carried on, in this case, through placing the user in a system of signs that come from high or legitimate culture. In design, the strategy is based on the legitimacy of the artistic forms employed, having their origins in the universe of high, rare, and privileged aesthetic experiences.

To make the "culture effect" succeed in conferring the sense of status, the legitimacy of the cultural opportunities must be acknowledged. Considering the scarce cultural capital of the user, the acknowledgment is made on the basis of historical associations, stereotypes, or mass media divulgations.

The upper-class public recognizes a style thanks to the principle of pertinence applied within its universe of analogies and differences of artistic production.

The principle of pertinence in a public having scarcer cultural capital, even if rich in economic capital, works when objects–signs are made recognizable by the context, amplified by hypervisibility, emphasized by stereotypes, and popularized by popularizers. It is "an aesthetic combining the unexpected with unconscious comparisons, imitations or past figures, the artist's fascination with the trivial and the commonplace, the culture industry in all its forms, the new themes engineered by films, comic strips, television, the prestige of 'unrefined art' and of consumerism."[17]

A classic example of this new strategy of architectural communication is Charles Moore's Piazza d'Italia in New Orleans. In this case eclectic references to tradition (Piazza d'Italia "conceives of history as a continuum of portable accessories"[18]) are made meaningful within a system of signs that produces an effect of high culture for a very broad audience.

The narrative and celebratory objective of Piazza d'Italia was achieved by combining styles and references. The classical environment of columns and fountains creates an exceptional and unique experience for the users, with the pretense of celebrating the Italians of the city and their national heritage.[19] The combined value of the building (in the case of Piazza d'Italia it was stressed that the cost was running at $1.65 million) and the artistic intent of its architect turned the design into an exceptional one, and the place into an artistic landmark that is worth visiting.

Artistic intention is an important factor that postmodern architecture shares with idealistic aesthetics. By means of this criterion, more appropriate for early-nineteenth-century European philosophers than for a contemporary *New York Times* critic, Paul Goldberger establishes a difference between two buildings similar in many other respects: the Piazza d'Italia in New Orleans and Caesar's Palace in Las Vegas.

> What differentiates something like this (Piazza d'Italia) from the cheap classical columns in front of a place like Caesar's Palace in Las Vegas? Both take classical forms and play with them, turning them into something easy and entertaining. The Piazza d'Italia was designed by a famous architect and is considered worthy of attention. Caesar's Palace is considered kitsch. Why? Part of the answer lies in the intent. In the New Orleans fountain, Charles Moore was operating on the level of exploring the meaning of classical elements; he wanted to make us think about what a classical column is and how it does its job, as well as entertain us. To do so, he created an intricate, subtle composition, full of wit and inventiveness.
>
> Caesar's Palace, on the other hand, is just a simple, quick-and-easy rip-off of classical elements in the belief that their very use, no matter how simplistic, connotes class.
>
> The Piazza d'Italia has no such naive aspiration toward "class." It is itself, its own thing, its own creation—like all real works of art.[20]

According to this analysis, Caesar's Palace was intended to create an environment in which, thanks to stylistic references to imperial Rome filtered through Cecil B. De Mille, an opulent movie set has been created in order to implement people's motivation to spend their money. Piazza d'Italia, on the contrary, can be defined as an artistic work in that it pretends to be as autonomous as any artistic output since Kant (*ars gratia artis*). Charles Moore's Piazza does not consciously intend to confer status. The same result of offering a social remuneration to its users is achieved in a more subtle way with the selective power of the "culture effect."

In this design strategy, everybody, both the upper classes and upwardly mobile middle class, must be able to act as or represent themselves as users of the cultural environment.

The high- and medium-social-status audience possesses above all a great deal of cultural capital even though their economic capital may not always be so high (among them would be the professional community of architects), and it is able to have an exclusive relationship with legitimate culture. The key to success to this social target is what we can call the Lichtenstein effect contained in the architectural design. This refers to the transformation by Roy Lichtenstein of a comic strip, blown up and isolated, in an autonomous painting. The comic becomes a work of art through irony and the so-called artistic distance from the object: In this way even what for most people can be trivial or banal is transformed into an artistic creation. The postmodern artistic work, either painting or building, communicates with double coding, as Umberto Eco and Charles Jencks define it.[21] A sign can be either a substantive communication to a part of the public or a quotation to another part (the one with greater cultural capital). This is the typical strategy of Robert Venturi, who assumes a culturally refined and competent audience as principal reference for his main street pop architecture. He ends up obtaining only a few important commissions, even though he is one of the most interesting, and quoted, architects on the contemporary American scene.

What Charles Moore calls the Brubeck effect acts as a means of communication to the same high social segment of the audience.

> The semipop musicians, like Dave Brubeck, do something very similar: they take ordinary themes and mess with them just enough, so that you can still recognize the themes, but you notice them for the first time, because something, maybe something awful, is happening to them, they're being kicked around in a way you didn't expect.[22]

The Brubeck effect works in an opposite but equally efficient way on lower classes. In this case single themes of the system are recognized and appreciated thanks to their belonging to the realm of everyday experience or "popular" art. By means of these themes the lower- and middle-class users can contact the art language as a whole and recognize it, without really understanding it. Thanks to the good cultural aspirations typical of rising classes, however, it will be taken for granted and accepted. From this point of view it can be said, following the musical analogy, that as far as a strategy of gratifying communication is concerned, Piazza d'Italia is closer to the Boston Pops, with its popularizing of classical "easy" music, than to Dave Brubeck.

On the contrary, for the educated upper class the aesthetic track is downward: from the artistic and autonomous coherence of the whole to the single themes, whatever they may be. The ruling classes are able, because of their aesthetic competence, to distance themselves from the object represented in the artwork. Even banal or trivial themes strategically contextualized and manipulated can become objects of artistic admiration. Because of these two levels of communication, postmodern context-specific designs, such as Robert Graham's Olympic Arch in Los Angeles, are accepted, albeit differently understood, by both segments of the public.

The postmodern museum applies to artistic consumerism in the same motivating approach that underlies the scenographic design of up-to-date shopping malls such as Georgetown Park in Washington D.C., where a "Victorian stage set" has been re-created to encourage sales.

The much-debated J. Paul Getty Museum in Malibu can stand as a good example of an appealing and motivating postmodern approach for a fine arts museum. It is a plastic and concrete copy of the Villa dei Papiri in Pompei which creates a coherent, though problematic, context of signs and stereotypes that permits the artistic character of the objects shown to be recognizable to all the users, even those who are not very familiar with exhibitions or museums. In homage to California one enters the villa straight from the parking garage. Furthermore, the Roman–Hollywood design, at odds even with the eclectic built landscape of southern California, emphasizes the exceptionality of the museum and of the experiences that can be had there in contrast to those of everyday life.

This same strategy, although realized by means of different architectural forms, is utilized in the museum for the De Menil Collection of symbolist and African art in Houston, Texas. In this case Renzo Piano intended his own museum-space to be as exceptional as the objects exhibited in it. Even while establishing connections with the wide-open spaces of Texas, the architect (designer also of Centre Pompidou in Paris) wanted to create something that could represent an "event" for people driving several hundred miles in order to enjoy an "unusual artistic experience."[23] In the case of the Getty Museum, the recognizability of a work of art and its intrinsic value is facilitated with the help of classical Roman stereotypes. In the De Menil Collection the artistic status and the exceptionality of the objects are revealed by their emphasized isolation and high visibility.

The traditional European museum, overcrowded with objects, demands from visitors the ability to "recognize," "identify," "sort," and "evaluate." That is, it requires the ability to use the principle of pertinence, which comes from adequate cultural capital. In the case of the De Menil Collection, the museum has practically been emptied: from the storage rooms (called the Treasure House) a few pieces at a time are brought to the public for a limited period "to avoid people having museum exhaustion."

This method of isolated display gives each object a character of exceptionality and creates artistic status for it even without the visitors' competence, thereby achieving the attention and the satisfaction from a large public.

The Architectural Scene

Postmodern architecture's intent is to give a large public the chance to take symbolic possession and control of the building and of the objects/signs contained within. The revolving cocktail lounges on top of the towers of Johnson's and Portman's hotels allow the symbolic reduction and appropriation of the surrounding landscape to be reduced in such a way that it may be "consumed."

Another major feature of postmodern design is the accentuation of the theatrical dimension of architecture even in the conditions of intimate use, required by the new relationship between people and built space. Architecture has always aimed, explicitly or implicitly, to create "theatrical" scenes and living shows. The problem today, different from that of the modern movement, is the production of a stage set effect even in the proximity of its use. For the rationalists design had a stagelike dimension to it, but the "show" was at a great distance and was extraneous to the user, who was never able to leave his role as spectator and become a protagonist on that stage. Even the architecture of Le Corbusier is "theatrical architecture seen from outside and at a distance."[24]

The tendency today is to create a stage set to be seen and experienced from close up and inside. The fundamental key to this communication strategy is participation, making an important shift from baroque scenographic architecture to contemporary neobaroque urban stage sets. The change is the new active role of the user–actor within the postmodern stage.

It has been stressed that baroque architecture can be defined as a spectacle from two different, yet complementary, points of view: as illusion and trickery and as a built living scene. If, as Mumford notes, the sensuality of baroque sculpture makes a statue similar to the human body, at the same time people stand for sculpture on the stage of the baroque built urban environment: It can be said that the people look like statues and take on their scenographic function.

In the baroque age—the great European Theater Age—the city and the great palaces become theaters themselves, with their scenes, stage sets, and perspectives. From the Trinità dei Monti steps and Piazza Sant'Ignazio in Rome to the Tuilleries in Paris, from the palaces of royalty in France and Austria, to the Counter-Reformation churches in Spain and Italy, the architectural environment becomes a stage.

If one considers how much a Roman baroque Counter-Reformation church was like a great stage, then the criticism of that masterpiece of cult-taken-as-a show, Johnson's Crystal Cathedral in Orange County, California, no longer seems as valid. From the point of view of communications strategies used, this postmodern church is much closer to the Roman seventeenth-century churches than to nearby Disneyland.

Dissimulation

The formal and social centrality of the scene constitutes one of the closest ties between baroque and postmodern neobaroque architecture. The urban baroque and neobaroque scene is made up of performances and trickery. Tricky theatrical baroque architecture does nothing but adopt and translate within the language of built forms the most shared ethical principle of that society: dissimulation.

Dissimulation is an indispensable criterion of behavior in a social system where political power is absolute and where forms and etiquette are commonly shared values, not only at the royal courts but in the whole society (the court society). Dissimulation, a Darwinian mode of survival within the social relations of the *ancien régime*, comes to be considered a virtue of the gentleman.

Della Dissimulazione Onesta (Of Honest Dissimulation) was the title of a noteworthy and widely known short Italian pedagogic treatise of the seventeenth century destined to educate noble youth.[25] According to its author, Torquato Accetto, a "virtuous" form of dissimulation is both possible and recommended in that "dissimulating not being anything other than a veil composed of honest darkness and violent respect; it does not create falsehood but allows truth to rest in order to show it at the proper time."[26]

Translating this ethic principle into baroque design means trickery and dissimulating *trompe l'oeil* architectures: "one simulates what it is not and dissimulates what it is." In baroque architecture just as in social relations, trickery and dissimulation are rules of the game and as such they are accepted. Therefore "we know that all that which is beautiful is nothing but a gentle dissimulation."[27]

The public demand for postmodern architecture to provide a social stage-show is different from that of the baroque era, insofar as today people not only want to look at the performance on the stage but want to play an active role in a play that is both real and subjective, true and deceiving.

The traditional baroque façade, which had become synonymous with deception and put apart by the modern movement in the name of purification and truth, has been reevaluated, symbolically overloaded, and stressed by postmodern architects.

The shift is from function to fiction. "We are phonies," says Johnson about himself and his own colleagues. His are not only visual tricks. Contemporary architectural theatrics turn shopping malls and suburban arcades into Victorian or bohemian environments. The transformation of a Holiday Inn into a noble English manor is a movie-set type of deception made possible by mass media culture. However, it deals not only with a neobaroque visual trick but also with a social trick, an honest and consensual dissimulation as foreseen in the treatise by Accetto.

The people of the affluent society ask to be protagonists in the trickery of the architectural scene insofar as trickery and dissimulation constitute the social scene.

The Goffman People

The people populating the postmodern stage sets of great hotels, restaurants, and holiday resorts are Erving Goffman's people. Goffman's reality of social order is the sum of images produced by social actors in the stage performances that they play for other social actors. That which Goffman's man asks from postmodern architecture is a stage set, real yet illusory at the same time, suited not so much to his current status as to his growing aspirations and desires. Life is not, as it was for the court society, a continuous parade, but it must still allow the chance for a parade and for conspicuous performances.

The Fontainebleau Hotel in Miami, chosen as a set for a James Bond film, was explicitly designed by Lapidus as a "movie set" ("All right, I'll design a fabulous movie set"[28]) in which the actors are the clients themselves. The restaurant, with its entrance on the same floor as the lobby, is reached by ascending a staircase, passing by a podium, and descending another staircase. It is not a folly of design, nor is it a neobaroque visual trick: it is a social trick.

> Why walk up, walk across and walk down? And yet no one has ever realized that there was absolutely no reason for it at all, except as a dramatic entrance to a restaurant. You walk out. You see everybody. Everybody sees you. You are dressed in your best bib and tucker. . . . I put those people on stage and they love it. . . . But, if anything, that entrance sums up my whole theory that

people love the drama and excitement of being part of the scene . . . being part of the show.[29]

The stage entrance's objective is to allow one to take part in a role played in front of people who are also spectators–actors and accomplices in this tricky social show. The social image these actors (especially those belonging to the shifting middle classes) want to create on the stage is intended to help them attain greater social status.

Once again Morris Lapidus, who has, as he stresses, a "tremendous indoctrination in merchandising," notes, "Some people demand the stage setting. They want background. They want status, just as the people in Seagram's achieved status because they sat against this impeccable historic background which none could question; they had instant status."[30]

It is the same strategy that, invented and brought to perfection in the 1950s on the Italian coast at Palinuro, made the Club Méditerranée a multinational corporation in the leisure business. The secret to its success was in having stimulated and utilized the ancient desire of men to become gods. In the Club-Village of Palinuro, French and German middle-class vacationers looked with admiration upon the beautiful young village staff black from the sun, making love and dancing, their *pareo* recalling Gauguin and dreamy southern islands. Everybody else was allowed to become a god himself. A simple act of will was enough to accomplish the metamorphosis. By overcoming the invisible barrier around the village square that separated actors from spectators, Olympians from mortals, everybody could turn himself into a god, and as such, into an object of envy and admiration.[31]

Political Consensus

Monumentality, scenography, eclectic use of traditional and pop culture, and participation make up the stylistic techniques through which postmodern architecture confronts the problem of legitimating power through built space. These make up the new vocabulary of signs for the ancient project of architecture to represent power and produce additional amounts of legitimation.

Monumentality has been a constant device in the architecture of power through the centuries, opposing ordinary people to the powerful, the ruling to the ruled, and making use of awe (the form of the castle) or admiration (the baroque or renaissance palace).[32] In both cases the common denominator is in that of making people acknowledge the power of the prince in the process of modifying nature and putting his stamp on the urban landscape.[33] This strategy continues today chiefly in developing countries. The maximum investments—economic and symbolic—are made for those buildings that hold power and represent new national identity. In Western societies the new princes are great corporations.

Today the growing problem for architecture stems from the fact that the built environment is no longer considered culturally obvious. That is to say that forms of built space are not taken for granted but are thematized and must elicit the users' consensus on what they are and on what they stand for.[34]

The difficulties in communicating effectively through architecture become much greater with increasing need for social and political consensus. It is the central demand of consensus that defines the architecture of Johnson and Portman as "humanistic" or that tells an entire generation of American architects that they have to design "with man in mind."

The corporate skyscraper of the post–Mies van der Rohe generation combines styles of late rationalistic functionalism with traditional and pop culture references. This monumentality tries to eliminate the gap between the building and the citizen by using and stressing a shared set of symbols and emphasizing participation in the name of common civic duties.

The skyscrapers of the great corporations, maximizing the use of urban land, reach upward to express the power and social presence of their owners. Differing in period and style, skyscrapers such as the Chrysler Building in New York, the John Hancock towers in Boston and Chicago, the Transamerica Pyramid in San Francisco—just to name a few—are vehicles of the same message of power: They are "pointing skyward optimistically like rising corporate spirits and rising corporate profits to be located in the clouds, a symbol of triumph. . . . The symbol of corporate power, the rising up of real estate, services, production, the variety of city functions integrated for a single end—a 'capital symbol'."[35] This is what Charles Jencks writes about the Transamerica Pyramid by Pereira, which, though much debated, has become the symbol of San Francisco itself.

Today this architectural display of power must be balanced by a new practical and symbolic relationship with the community. Contrast, embedded in the concept itself of monumentality, must be achieved through a system of values and references that can be understood and, above all, politically accepted by people. The old strategies of inducing awe and admiration now merge with ones founded on identity and participation, on a quest for people's consensus by means of shared sets of values and forms and using the old American political theme of community participation.

The effect desired by postmodern architects is similar to what Vincent Scully, the Yale architectural historian, found in Michael Graves' monumental Portland Building (where classical forms merge with "supersigns" of authority, allusions to ancient Egypt coexist with quotations of Grand Central Terminal in New York): "a totally unexpected cultural assurance. One is at home there".

Functional and formal rigorous musts of the modern movement are postponed by the consumers' need to feel at home. Architecture no longer expresses the need for purification; it reflects the contemporary hedonistic demand for day-by-day well-being. To recover the users' confidence, postmodern architects seek a reconciliation with history and tradition after a rationalistic break with the past and stress on an ergonomic and purifying future.

In the same project of recovering traditional and widely shared cultural models of everyday life, the functionality of forms and the massive vertical monumentality of Johnson's AT&T building in Manhattan are balanced by the allusion to a Chippendale highboy. Recalling ecological values and making the most of nature are themes used in the architectural corporate image of Chemical Bank in their 277 Park Avenue tower, where, in its 1981 renovation, the lobby was transformed into a tropical greenhouse.

The IDS Building in Minneapolis, the Citicorp Center, and the IBM Building in Manhattan establish new practical and symbolic ties with the city and its people by re-creating the town square inside the building, which is "the living room of the city," to use an expression by Johnson, in order to allow everybody to use the building not as a customer but as a citizen.

The design has the community enter the corporation building both symbolically and practically in order to stress the relevance of the corporation to the public and thus gain consensus through the building for the corporation itself. From this point of view these "humanistic" buildings are, in the realm of built space, the equivalent to cultural foundations funded by corporations. The building operates on an image strategy insofar as the internal square, with its greenhouse, cafeteria, church, and shopping mall, allows no more real participation in IBM than that of a humble shareholder in IBM management. This gap between images and reality is filled once again by dissimulation, the founding principle of postmodern architectural communication.

Notes

1. Giandomenico Amendola, *Uomini e Case: I Presupposti Sociologici della Progettazione Architetonica* (Bari: Dedalo, 1984), pp. 126–33.
2. The operational translation of this "doubt about intention" is the distinction by Herbert Gans between potential and effectual space. The former—the space of the design—is the architect's intention; the latter—the real space produced by people who use the de-

signed environment—is the effect of the action, which often does not correspond to the architect's intention. Herbert J. Gans, *Power, People and Plans* (New York: Pelican Books, 1968, 1972).

3. Charles Jencks, *The Language of Post Modern Architecture* (London: Academy Editions, 1977), p. 9.

4. Jurgen Habermas, "Moderne und Postmoderne Architektur," *Arch +*. (February 1982):54–59.

5. Francesco Dal Co, *Abitare nel Moderno* (Bari: Laterza, 1982), p. 74.

6. Robert Venturi, *Complexity and Contradiction in Architecture*, New York: The Museum of Modern Art, 1966, p. 16.

7. Roland Barthes, *Mythologies* (Paris: Seuil, 1957).

8. Sigfried Giedion, *Space, Time and Architecture* (Cambridge: Harvard University Press, 1941), p. 89.

9. Gillo Dorfles, *Architetture Ambigue: Dal Neobarocco al Postmoderno* (Bari: Dedalo, 1984), p. 22.

10. "Houses take different names according to the dwellers' status. It is then called maison the house of a bourgeois; hôtel the house of an aristocrat; palais the house of a prince or of a king. . . . Therefore, nobody else whatever his social rank could be but those people, is allowed to engrave on the entrance door of his house, the word 'palais'. Even the hôtels—*les demeurs des grandsigneurs*—must be different from the royal palaces. Their decor requires a beauty adequate to the rank and to the birth of the people that had them built. Although they must not show the same magnificence that is reserved for the king's palace." "In short . . . people who had not the same social rank had to live in houses whose structure could show superiority or inferiority of different status orders." In *Encyclopédie, planches*, Vol. II, part V, quoted by Norbert Elias, *Die Hofische Gesellshaft* (Darstad und Neuwied: Luchterhan Verlag, 1975), p. 38.

11. Gaston Bachelard, *La Philosophie du Non*, 2d ed. (Paris: PUF, 1949).

12. Elias, *Die Hofische Gesellschaft*, p. 38.

13. Lewis Mumford, *The City in History* (Orlando, Fla.: Harcourt Brace Jovanovich, 1961), Italian trans., pp. 435–68.

14. Pierre Bourdieu, *La Distinction* (Paris: Les Editions de Minuit, 1979).

15. Ibid., p. 39.

16. Morris Lapidus in John W. Cook and Heinrich Klotz, *Conversations with Architects* (New York: Praeger, 1973), p. 152.

17. Jean Duvignaud, *Sociologie de l'Art* (Paris: PUF, 1984), p. 167.

18. Henrich Klotz, *Postmodern Visions* (New York: Abbeville Press, 1985), p. 180.

19. Piazza d'Italia represents a working example of cultural hybridization and popularization where the eclecticism is all contained inside the same monument with no aim other than to make it appealing and gratifying to the public. It is thus different from the nineteenth-century eclectic historicism that intended with a particular style to recall a specific historical epoch and thus emphasize the function of the building (i.e., all the public buildings of Vienna Ring are of different meaningful styles). It is also different from the postmodern context-specific architecture that makes form, styles, decors of the building recall the destination of the building.

20. Paul Goldberger, *On the Rise: Architecture and Design in a Post Modern Age* (New York: Times Books, 1983), p. 136.

21. Charles Jencks, *What Is Post-Modernism?* (London: Academy Editions, 1986), pp. 18–19.

22. Charles Moore, in Cook and Klotz, *Conversations with Architects*, p. 236.

23. Quotations come from a conversation with Renzo Piano in 1985.

24. Henri Raymond, *L'Architecture, les Aventures Spatiales de la Raison* (Paris: Centre Georges Pompidou, 1984), p. 184.

25. This treatise by Torquato Accetto was published in Naples in 1641. It was published

(Bari: Laterza Editore) again by the philosopher Benedetto Croce in 1928 during the fascist period. Quotations are from a reprint (Rome: Marginalia, 1982).

26. Ibid., pp. 18–19.
27. Ibid., p. 29.
28. Morris Lapidus, in Cook and Klotz, *Conversations with Architects*, p. 156.
29. Ibid., p. 157.
30. Ibid., p. 173.
31. Henri Raymond, "Recherches sur un Village de Vacances," *Revue Francaise de Sociologie* 3 (I) (1960):323–33.
32. Harold D. Lasswell, The *Signature of Power: Buildings, Communication and Strategy* (New Brunswick: Transaction Books, 1979); Amendola, *Uomini e Case*, pp. 137–67.
33. Luigi Firpo, *La Città Ideale del Rinascimento* (Torino: Utet, 1975).
34. Giandomenico Amendola, *Sense and Identity in Urban Change*. V International Conference of Euro-Arab Social Research Group, Cairo 1983, mimeo.
35. Charles Jencks, *Skyscrapers-Skycities* (New York: Rizzoli, 1980), p. 15.

13

Architecture Invents New People

ROGER MONTGOMERY

"Modern Architecture died in St. Louis, Missouri on July 15, 1972 at 3:32 P.M. (or thereabouts)," wrote Charles Jencks, the critic and historian, in the opening chapter of his *Language of Post-Modern Architecture*, "when the infamous Pruitt-Igoe scheme, or rather several of its slab blocks, were given the *coup de grâce* by dynamite. Previously it had been vandalized, mutilated, and defaced by its black inhabitants, and although millions of dollars were pumped back, trying to keep it alive (fixing broken elevators, repairing smashed windows, repainting), it was finally put out of its misery. Boom, boom, boom. . . ."[1] This dramatic passage expressively symbolizes an important transformation in architecture in the United States and much of the rest of the Western world as well. The transformation centered on the discovery of, and central place given to, the occupants or "users"[2] of buildings. These occupants and users constitute the architects' "new people"[3] of this essay. The pages that follow explore how these new people came into being and how the transformations that centered on them have affected architecture practice and ideas.

Jencks, in the passage quoted earlier, continues, "Pruitt-Igoe was constructed according to the most progressive ideals of CIAM (the Congress of International Modern Architects) and it won an award from the American Institute of Architects when it was designed in 1951." Actually, as Jencks tells the story, he is wrong on almost every detail from the date of the dynamiting to the mistaken assertion that the project design won an award.[4] But quibbling over the veracity of Jencks' story misses the reason for recalling this flawed passage. Its value instead lies in its symbolization of change in the architecture design process. Jencks's assertion that not only did Pruitt-Igoe's "black inhabitants" bring down that infamous project, but their actions also marked the collapse of the dominant architectural paradigm of the twentieth century, elevates the occupants or new people to a pivotal position in design practice and theory. This central position of occupants and users actually represents a direct outgrowth of "functionalism," a central tenet of the modern movement itself and almost synonymous with it, not something new. The first part of this essay sketches in this part of the transformation.

The definition of *modern movement* as "functionalism" opened the way in both practice and theory to a fruitful interaction between architecture and social science. For twenty years or more before Jencks published his words on Pruitt-Igoe, a small number of social and behavioral scientists along with a few architects committed to a particular reading of modern architecture had worked to articulate a new model or paradigm of the architecture process. Based on the conflation of social science and architecture around functionalism, this new model began with the human beings who occupied buildings, the users. The inventors of the paradigm saw in its use the opportunity to make far better buildings in terms of satisfying the users (which is what they defined as better). In defining architecture this way the occupants and users became the first causes around whom the whole architecture process revolved. They became, in short, the architects' *new* people, people quite distinct in their objective clarity from the subjective homunculus of the architects' *old* people or tacit, imagined occupants of the world before the conflation. By studying these new people and using the findings of such study in the environmental design[5] process, buildings that better served their users and occupants would result. The second and third parts of this essay treat the rise and definition of the new paradigm.

Under the unique historical conditions that obtained during the late 1960s and early 1970s, the new paradigm for the architecture design process seemed well on the way to becoming the dominant mode of work. Study of the new people became an architecture specialty trade in and of itself, a defined subprofession with a widely understood nomenclature that included the use of the term *person–environment relations* to denote the new field of activity.[6] Especially in the world of professional education the paradigm seemed unassailable. Yet what had seemed so certain a decade earlier had by the end of the 1970s itself become transformed. The fourth part of this paper treats this highwater point and the beginnings of these most recent changes.

How did this new paradigm emerge? Does it work? What unanticipated results have occurred? The final part of this essay turns to this last question. In so doing, it suggests an alternative interpretation of the new paradigm, one that sees the new approach not as a better way to do architecture, but rather as part of the societal processes that continuously elaborate the division of labor. This alternative view associates the discovery, or, more accurately, invention, of the new people with new kinds of architecture labor and with ongoing elaborations of the architecture production process. Thus, viewed from the perspective of those who had created the new paradigm, the invention had unanticipated results. Seen from a wider, societal perspective, the process appears part of a larger project concerned with extending the division of labor in postindustrial society and increasing the relative role of professional services required to construct the built environment. The invention of architects' new people paralleled, legitimated, and even helped create not new architecture but new jobs.

The Modern Movement and Functionalism

A properly detailed exegesis of modern architecture is impossible in this short essay. If it were it would include a thorough critical interpretation of functionalism as a central tenet of the movement. Suffice it here to note that the two terms, *modern architecture* and *functionalism*, came to be nearly interchangeable. Two reasons for this are frequently noted: first, the audience has tended to take the pioneer modern architects' polemics at face value. For example, architects, critics, and the general public tended to understand Louis Sullivan's "form follows function" and Le Corbusier's "a house is a machine for living" to mean what they said in everyday terms.

The second but equally important reason: early, nonarchitect enthusiasts or apologists interpreted the new architecture in narrowly functionalist terms. Henry-Russell Hitchcock, mid-twentieth-century America's most important interpreter of the movement, played a central role in defining the enterprise interchangeably as modern architecture or functionalism. In 1932, in collaboration with Philip Johnson, Hitchcock wrote, "it is an absurdity to talk about the modern style in terms of aesthetics. . . . If a building provides adequately, completely, and without compromise for its purpose it is . . . a good building, regardless of its appearance."[7] This kind of talk and the theory implicit in it became so commonplace during the 1930s that *functional* became a synonym for *modern*. To cite another example, J. M. Richard, the British editor and critic, in his influential 1940 text *An Introduction to Modern Architecture*, argued that "good architecture is produced automatically by strict attention to utility, economy, and other practical considerations."[8]

Behind the discourse of critics and polemics of architects lie critical material dimensions of the building economy. Reyner Banham, historian and interpreter of the modern movement, connects the rise of functionalism to the political economy within which the modern architects worked. He observed of the mid-1930s architecture scene,

With the International Style outlawed in Germany and Russia, and crippled economically in France, the style and its friends were fighting for a toehold in politically-suspicious Italy, aesthetically-indifferent England, and depression-stunned America. Under these circumstances it was better to advocate or defend the new architecture on logical and economic grounds than on grounds of aesthetics or symbolisms that might stir nothing but hostility.[9]

Only recently have scholars begun to unravel some of the specific connections between the modern movement and the political economy of building and urban development. For example, Gerhard Fehl has uncovered connections between dominant sectors of the German construction materials industry and the distinctively different approaches to mass housing design associated with Walter Gropius in Berlin and Ernst May in Frankfurt.[10] Much remains to be done to complete the case for a structural understanding of the rise of functionalism, but the outlines of the story seem clear enough to support the argument presented in this essay.

On the ideological level another aspect of the modern movement played an important role in bringing architecture into contact with social science. During the period when the movement arose, Western culture widely and uncritically accepted the notion that both contemporary art and contemporary science expressed or revealed the same underlying *zeitgeist*. This meant that art and architecture held a strong positive, ideological commitment to rationalism and science. In its extreme form this view had cubism in painting prefiguring quantum mechanics in physics. The everyday variety of this ideology posited modern architecture as expressive of the "machine age" or industrial revolution. More generally, this implies a social functionalism that permits the "concretization of social institutions and values characteristic of particular cultures or eras,"[11] and a cultural functionalism that involves the "concretization of universal values and subconscious structures of spatial and psychological orientation." Seen on this level of generality, architecture and social science have very similar missions and the former seems almost an applied branch of the latter.

Another ideological connection lay in the mostly European alliance between the new architecture and radical politics. Perhaps inevitably, the turn-of-the-century artistic revolutions converged in people's minds with the vigorous European socialist and communist movements. Artists became part of the vanguard not only in art but also by virtue of analogic thinking in social and political affairs. The high point in this convergence occurred in the early postrevolutionary Soviet Union when a close identity developed between the Constructivist and Suprematist movements in art and architecture and the Bolsheviks. More important in terms of the evolution of mainstream modern architecture was the long romance between urban social democratic governments in Western Europe and functionalist design. Monuments of this era include Dr. May's famous *siedlungen* (settlements or suburbs) in Frankfurt am Main, Berlin's broad social housing program during the Weimar Republic, and, of course, Vienna's housing projects, including Karl Marx Hof. The pinnacles of this alliance were reached in the Scandinavian countries and Holland, where, under social democratic regimes, architecture truly represented the concretization of socialist institutions and values.

But how did the connection between architecture and socialism relate to the conflation between social science and architecture? The answer lies in the socialist demand that art must serve people, hence social realism in painting and literature, social housing in architecture. An architecture in the service of social housing placed the emphasis in the design process on the needs of the working-class occupants. This in turn forged links, at least in theory, between architects and social or behavioral scientists in order to understand those needs better. Dr. May's use of controlled social scientific methods to achieve an ideal ergonomic kitchen design for use in Frankfurt's social housing program represents this connection in practice.

Given the redefinition of architecture as primarily concerned with a building's purpose, and given an ideological commitment to rationalism and socialism, architects had a clear motive for looking to social science to tell them what they needed to know. The best illustrations of this appear in the work of another vanguard modernist, Walter Gropius. In 1929 he wrote on the "sociological premises" for urban housing and argued that, "The sociological facts must first be clarified in order that the ideal minimum of a life necessity, the dwelling . . . may be found . . . an entirely new formulation is required, based on a knowledge of the natural and sociological minimum requirements, unobscured by the veil of traditionally imagined historical needs."[12] Though in the historical situation neither Gropius nor other important pioneers in modern architecture actually followed up on this and included social research or sociologists or other behavioral scientists in the building process, the idea had been articulated. Although this part of the story had taken place in Europe, the next chapter would unfold across the Atlantic.

Social Science and Functionalism in Architecture

From the social science side the conflation with architecture can be dated by the appearance of a single empirical study in the United States. Research carried out in the 1940s by Leon Festinger and his associates at MIT examined the relationship between the physical layout of dwelling entries and face-to-face interaction patterns in a married student housing project. Festinger's study marked the start of a new social science concern with the built environment. [13] It helped shape the consciousness of a group of workers who, during the 1950s and early 1960s, set forth the empirical and theoretical basis for a new field of person–environment relations. Festinger and his colleagues showed that friendship patterns among people correlated closely with the physical relationships and proximity of their apartment front doors. Almost immediately some architects and a few tentative interdisciplinary types interpreted these findings to mean that by consciously designing the relative positions among front doors of a housing project, the designer could determine the friendship patterns among the occupants of the project. [14]

To make this claim for the Festinger study as a catalyst is not to claim that no one previously had made the connections between the built environment and behavior. Indeed they had. Actually the study revived a long-standing environmental determinist perspective. More than a century earlier, reformist public health efforts to mitigate slum conditions in England and later in the United States connected behavior to environment. Drawing on the model provided by early scientific studies of communicable disease, primitive social science made uncritical comparisons of behavior patterns with environmental conditions. For instance, data on the spatial distribution of poor housing conditions viewed together with the spatial incidence of alcoholism were used to argue that slums caused drinking. Such numbers typified what sociologist Robert Merton pejoratively called "social bookkeeping"; they said nothing about causality. Drinking could as well cause slum housing. But such an alternative reading did not fit the dominant ideology, and environmental determinism won the day.

American social science gave further strength to environmental determinism early in the twentieth century with the rise of the Chicago school of sociology. The perspective of this group, called human ecology, viewed social relations as generated by territoriality and thus dependent upon the physical environment as well as on social, psychological, and linguistic dimensions. [15] With the complementary zone hypothesis of the "natural" life and death of neighborhoods, this sociology fit precisely with the emerging dynamics of the urban real estate market in the mature U.S. economy. Human ecology and the zone model provided the legitimating rationale for slum clearance and urban renewal. By the 1930s, this use of social science brought it into the world of active urban policy formation, and, in the process, into direct connection with the social housing wing of architecture. [16] Thus, this most significant school of American social science helped pave the way to the conflation with architecture.

Concurrently in the 1930s, some studies began to raise questions about the validity of the public health, environmental determinist model. "Indeed," Merton observed, "one of the very few studies attempting to locate the distinctive part played by substandard dwellings as such holds it 'unmistakable that there is no relationship between bad housing in its physical aspects and juvenile delinquency as revealed by court records.' "[17] The debate continued between skeptical social scientists in the Merton tradition and the more applied workers with their notions of environmental determinism. Twenty years later, a team of public health researchers under D. M. Wilner undertook a monumental, longitudinal, public health study in Baltimore. The research compared an experimental population of former slum dwellers now housed well in brand-new dwellings with a matched control population of slum dwellers.[18] The results indicated that improved housing had little or no effect on critical measures of physical and mental health. However, the issues raised by Wilner's group and others whose empirical work questioned environmental determinism did not deflect the trajectory set by Festinger. In the course of the next decade a notable literature on person–environment relations began to appear. Among the better-known contributors were anthropologists Anthony F. C. Wallace and Edward T. Hall, sociologists Leo Kuper and William H. Whyte, Jr., and from a very different side of that discipline Erving Goffman, psychologist Roger Barker, and social psychologist Robert Sommer.[19]

Wallace's study, an assessment of high-rise public housing for large, inner-city families, can represent the work. A clear functionalist conception of architecture underlay the research. Wallace asked whether high buildings or low ones best served the needs of public housing occupants, mainly mothers and their young children. He concluded that family life suffered in elevator-access, high apartments as compared with walk-up units and row houses. Such focused, applied research helped to build a cadre of social scientists committed to the study of relationships between the physical environment and human behavior. Through the 1950s and 1960s work continued. With the publication of Hall's *The Hidden Dimension* in 1966, Barker's *Environmental Psychology in 1968*, and Sommer's *Personal Space* in 1969 the social scientists had articulated a first complete formulation of the new field of person–environment relations.[20]

In the early 1950s, few people from architecture had the time or interest to explore the possibilities opened up by the Festinger study. On one level this may be explained as the normal small beginning of the innovation diffusion process. On another level material conditions may have blocked change. A building boom had begun after long years of depression and war had sharply restricted the training and entry of new people into the field. The 1950s proved to be a period of more than full employment for the then limited number of architects. Scarce talent had more than it could do simply getting projects built. New dimensions of practice, especially ones that required new training and a new professional outlook such as the evolving social science connection would require, could not elicit much interest. For the most part the new work by the social scientists interested in the built environment simply did not reach the architects.

However, a very few architects began to discover Festinger and the new person–environment work. They then made tentative efforts to build their side of the bridge between architecture and social science. Representative of this coming together, several young University of California, Berkeley, architecture faculty members undertook to link social research to the design of new dormitories on their campus. This vanguard work under the leadership of Sim van der Ryn and Murray Silverstein, published in 1967 as *Dorms at Berkeley*,[21] nicely represents the template that over the next decade guided much of the work in the new person–environment relations field. Embedded in this project and a few other contemporary pioneering efforts, such as those of architects Neal Deasy and Louis Sauer,[22] lies the model that directed research and practice and that became the field's dominant ideology.

One caveat before turning to a brief examination of the model. Despite the tone of the argument in this essay, which may suggest in places that ideology and scholarship powered history, changing material circumstances largely drove events. Even the intellectual history of these events demonstrates this. Why, for example, did a comparative flood of publications exemplified by fat, edited collections by academics like Proshansky, Michelson, and Gutman[23] appear in the wake of the Hall, Barker, and Sommer volumes? These works reprinted the substantial scholarship on person–environment relations produced over the preceding ten or fifteen years. Their appearance testified not only to a sudden increase of interest in the field at the level of ideas, but to new conditions in the architecture economy. These new conditions made publication profitable and therefore possible. The acute labor shortage of the 1950s had ended, and new schools had greatly increased the production of trained architects. Both the new schools and the new graduates formed a market for new books. The architecture profession had entered an expansionist phase that included the beginnings of a new subprofession in person–environment relations. These changes are treated later in this essay, after the more detailed examination of the new model or paradigm.

Social Factors in Architecture: The Ideal Model and the Invention of Architects' New People

In the decade marked by the first writings of Hall and Sommer, the role of social science knowledge in architecture practice seemed deceptively obvious to a growing band of designers and social scientists. By incorporating social science findings on the needs of building users and occupants into the architectural design process, the resulting buildings would function better for these users and occupants.

The expected role for social science in architecture, and the imputed systematic clarity of design practice incorporating social science inputs, can be represented by a simple, deterministic, and positive model.[24] This model idealized practice as a linear process beginning with knowledge and research, followed by a programming stage that specified the social knowledge to be incorporated in the building, then a final design stage incorporating that social knowledge. The presumption was that, if followed, this process would result in buildings that would be as well fitted as possible to the needs of the people who occupied and used them. An additional stage of follow-up or "postoccupancy" study to ascertain the quality of the fit became an essential final step. This last step completed the circle by providing the source of new social science knowledge to inform the design processes for the next round of projects. Figure 13.1. diagrams this ideal model.

Figure 13.1 The new people model or paradigm of the design process

By the end of the 1960s this model seemed so obvious and so persuasive that many felt it ought to prove self-implementing. As a paradigm for practice it identified for professionals from both architecture and social science a particular group of people who, according to the paradigm, played a transcendently important role in the design process. These key people were, of course, the users and occupants, not the architects and patrons who historically had held the central roles in the design process. These newly central people, users and occupants, identified and researched with the tools of social science, became the "architects' new people." They replaced not only the architects and the patrons, but the architects' uncritically accepted tacit "old" people.

Parenthetically, note that the papers in this volume, though studies of what this writer calls architects' "old" people, could not have been written before the invention of the new people. Prior to the definition of users and occupants as proper subjects for architectural and social scientific inquiry and concern, architects' people had no real existence. Outside of architecture and social science, other kinds of practitioners and scholars had often been interested in building inhabitants (archeologists, for instance, and tenement house inspectors). But only after social science and architecture had come together under the material and ideological conditions of the modern movement and Chicago sociology did the inhabitants of buildings become problematic. This illustrates a central point of this essay: without the discovery of the new people Jencks' commentary would have been impossible, as would this book and the careers its publication represents. Architects' new people, and the new design model implied by the phrase, have created new kinds of discourse, and new discourse represents new kinds of work—but more on that later.

The New Paradigm in Practice

To return to the main trajectory of this account, in the late 1960s, within the United States and Western Europe, crisis and upheaval set the political-–economic context of architecture and the newly emerging field of person–environment relations. This turmoil gave special urgency to the adoption of the new paradigm. Especially in the universities it seemed a historical necessity. The Paris uprising of 1968, in which the architecture students of the Ecole de Beaux Arts played an important vanguard role, symbolized for many the end of architecture as art and the arrival of architecture as a field interested in making places for people—meaning users and occupants. In the United States, where most of the first round of work on the new model had been done, two events, one on each coast, paralleled the Paris revolt. In New York City, architecture students led the sometimes violent sit-in strike intended to secure the use of Columbia University's new gym for the surrounding Harlem community. The next year saw even more violent events at the University of California, Berkeley, around a vacant block acquired for new dorms but at the time (and to this day) occupied by squatters from the "community," turf popularly known as "People's Park." These actions and others like them, though certainly mainly the products of an era marked by almost ubiquitous grassroots activism around issues of racism, war, imperialism, and what Manuel Castells has taught us to call collective consumption, had a special meaning for architects and the person–environment relations oriented social scientists. These events seemed proof positive of the dangers in following old notions of architecture as art and technique unconcerned with users and occupants. Paris and Columbia, 1968, and Berkeley, 1969, constituted the most obvious signs that the new model's time had arrived. Putting the new model to use became urgent in an effort to maintain social peace.

In the calmer setting of academic and vanguard architecture discourse, two important scholarly studies of this era illuminated a path for this effort. In the context of 1960s social upheaval, their new paradigm critiques of mainstream modernist social housing helped people understand the uprisings and provided a clear direction for design practice in the years ahead. In 1969 Philippe Boudon's study of Le Corbusier's Pessac housing created an enormous impact by indicating that the work of the most important modern master did not satisfy its users and occupants.[25] The implication was clear: Without social science and the new paradigm, functionalist motives, even in the hands of the most accomplished and acknowledged master, escaped realization. A parallel work, focused on a local rather than international modernist star, and involving a systematic empiricist research campaign, appeared in Clare Cooper's study of Easter Hill Village in Richmond, California.[26] These studies proved landmarks. They were among the first postoccupancy studies of the sort called for in the paradigm. Though they developed a strong case that the studied projects had not served their occupants well, they also gave equally strong indications for doing better in the future. In the context of the social upheavals, studies such as those of Cooper and Boudon mediated between design practice, social research, and the world they served. These studies provided concrete indications for the way the new model would work in practice.

Understandably, almost overnight important parts of the architecture establishment embraced the new paradigm. In the schools the change seemed complete. Everywhere new courses were taught that attempted the social science–architecture marriage. Even the prestigious eastern design schools at Harvard and MIT employed social scientists on their faculties to teach "social factors" to a voracious Vietnam War era student body. The architecture faculty at Berkeley, which had been doing pioneering person–environment relations work since early in the 1960s, landed a major federal grant to set up a doctoral program to train cross-disciplinary researchers.[27] Practitioners at all levels in firms of all sizes began to talk about user needs and employ social research consultants. For the time being, even the major corporate practices such as world-famous SOM (Skidmore, Owings, and Merrill) at least briefly joined the bandwagon. By the early 1970s, the architects' new people had arrived. To illustrate this, in what must be among the most bizarre events in the tragic history of Pruitt-Igoe, the Chicago division of SOM undertook in 1972 (the year of the dynamiting) to comprehensively redesign this ill-starred project, and thus presumably turn vandals into model citizens (blacks into whites?) through the beneficent influence of design informed by social science.[28] It seemed perfectly clear that the architect was emerging as an important species of behavior modification specialist—virtually an applied social scientist—adept at designing environments that would engender desired behaviors among occupants and users, and social peace in the ghettos.

Oscar Newman produced in his *Defensible Space* the seminal representation of the new paradigm in practice.[29] Published in the year of the dynamiting at Pruitt-Igoe and the SOM redesign effort, the book argued that the high rates of violence and vandalism in such public housing projects occurred because designers had not taken into account the occupant's territorial predispositions. Newman argued, and offered some data in support, that if public housing were designed so as to give rein to people's territorial tendencies, projects would be self-policing. Immediately on publication the book became a best seller among architecture texts. Newman had demonstrated the case for the new paradigm. In the language of the new model, he had demonstrated the failure of the old way of designing for tacit people. He showed successful design resulting from the use of science to articulate the new people. The fact that the Nixon Department of Justice paid for Newman's research and publications does not so much call them into question as provide splendid evidence for the material conditions surrounding the rise of the new paradigm.

During the 1970s many architecture professionals joined the movement. An even larger number of the people on the academic side of architecture, especially advanced graduate students and young faculty, formed the advocate cadre for the new model. Many indications testify to this and to the rapid spread of the ideas abroad, especially to Western Europe and the United Kingdom. Institutions for research and education were formed and associations organized, most important of which was the Environmental Design Research Association, or EDRA. Membership and annual meeting attendance went from dozens to hundreds during the decade. Journals and other serial publications started, such as the successful and well-known *Environment and Behavior*.

As the decade advanced, problems with the use of the new model dampened some of the enthusiasm for it. Although Newman's reconstructed public housing projects may have experienced less vandalism, ghetto violence continued, though mostly contained within the ghetto. Cities as a whole seemed quieter; at least the fear of social upheaval diminished. A buoyant building economy emerged in the last half of the 1970s. With it came new opportunities for architects to assume once again their traditional roles as "decorators of the *milieux* of elites," in the unforgettable phrase of C. Wright Mills. In the flush of new opportunities and with the decline of social pacification motives, the rise of postmodernist concerns with the aesthetics of style, corporate identity packaging, and elite environments should come as no surprise.

These are the conditions behind the postmodernism Jencks set out to legitimize in invoking Pruitt-Igoe. Modern architecture did not collapse, as Jencks claimed; it evolved into the postmodernism he presents as its dialectical opponent. First came the identification of modernism with functionalism, then the conflation with social science, the rise of the new model, and its use in interpreting the Pruitt-Igoe story. In invoking that troubled project, Jencks demonstrated ironically that the genealogy of postmodernism includes both CIAM's functionalism and Newman's user needs.

Throughout the 1970s, as postmodernism began its rise, the new subdivision of architecture practice that drew on the new paradigm flourished. The next section of this chapter discusses some aspects of the paradigm's efflorescence. To conclude this section, and to indicate the paradigm's maturity two recent treatises on architecture merit attention. John Zeisel has brought together in a single volume, *Inquiry by Design*, the methodological practices and insights created during three decades of development and use of the new paradigm.[30] More recently, Jon Lang has produced an enormously ambitious effort to synthesize within the framework of the new paradigm not only rules for practice in the present world but the whole corpus of architecture and architecture-related history and theory, from Vitruvius in ancient Rome to postmodernism in the present-day United States. The full title nicely indicates the line of argument: *Creating Architectural Theory: The Role of the Behavioral Sciences in Environmental Design*.[31] These recent books, with their doctrinaire adherence to the new model, represent the paradigm in the academy, not in practice. This raises the question, "How have the new people, so well represented in doctrine and scholarship, fared in the daily world of architecture work?"

New People and the New Division of Labor[32]

Critical changes in the American construction economy have occurred over the years since the person–environment relations people first articulated their new paradigm. In terms of this chapter the most important among these changes are those affecting the relationships between architects as providers of professional labor inputs to the building construction process and that construction economy. From this perspective two broad trends demand attention: first, the increase in the producer services component of the construction process and the related drive toward ever greater professionalization of those services[33]; second, the increasingly fine scale and specialized division of labor that directly has made room for the person–environment relations subfield.[34] As part of these secular changes the numbers of architects employed in the construction economy have greatly increased in relation to the actual amount of construction.[35] This in turn means either that the productivity of the architecture professional has dropped significantly (figuratively, architects have slowed down, so that it takes more of them to screw in the same light bulb) or that a given amount of construction requires more architecture inputs (again figuratively, the light bulb is getting more complex so it requires more people to install it). A third alternative exists: architects may be getting relatively cheaper than other construction inputs, so that the production function has changed and architects are being used in place of other inputs.[36]

Before turning to the relationships between these changes and the new architects' people in practice, it may be useful to illustrate some of the kinds of forces behind these secular trends. Perhaps the most dominant of these forces comes from an increasingly rationalized and globalized financial sector. This dominance has the effect of increasing the relative importance of process dimensions in architecture as against the built product. For example, this occurs as part of the financial sector's demand for more technical inputs to minimize risk and ensure marketability or consumer acceptance—and, in the settings of financial institutions, to communicate, to make evaluations, and to do the myriad processing common to bureaucracies. These newly important dimensions of the building process create new jobs for professionals. To take a concrete case from the financial sector, under the influence of bond underwriters, a public hospital becomes much less a civic monument and more a tool for the provision of health services; in the most extreme case, it becomes simply an adjunct of the financial system that assembles and allocates capital for public works. The effects on design work from such changes in the construction economy seems unmistakable: the artist-architect has less importance, as do experts in materials and construction detailing, or building structures and foundations. At the same time, the new paradigm architectural-programmer who has specialized in the spatial requirements of modern health care technology rises to new heights of importance. The specialists who mediate between the construction and financial systems to assure the credit-worthiness of the project gain enormous authority. In addition to the ever-present bond attorneys and underwriters, the public hospital development process includes environmental design people to evaluate physical design from a financial risk standpoint, much of it a species of market research otherwise known as user needs analysis.

Equally important changes in the political context of construction have had critical impacts in accelerating the division of labor within the construction industry and increasing the demand for producer services inputs, including architecture. These aspects of the changes have contributed to the rise of person–environment relations as a professional field. Political interests in construction activity and political power over construction decisions have become widely diffused through the society. In the United States today, almost nothing gets designed and built without multiple levels of political involvement, both formal and informal. The Byzantine apparatus of public regulation over land and development is perhaps only the most obvious manifestation of this change. Regulation has created very substantial numbers of middle-level professional jobs. All sides in the process have to employ experts both to manage the bargaining process and to provide technical analysis and representation. Negotiation, and the kinds of expertise it requires, draws far more on the person–environment relations model of architecture than on the more traditional perspectives.

To take a different example, employee participation in workplace design marks another type of change in the political context of environmental design. Higher levels of participation in design decision making, usually emphasizing participation by users and occupants, thread through the entire building economy today. Planning for workplaces and public use facilities as a matter of course includes representation of the future occupants or users on the client negotiating team. The architect must serve as a mediator, translator, and manager of these group processes in addition to serving as technical expert on occupant and user needs. These changes shift the relative importance from the artistic dimensions and material technology of architecture to various types of people work for which the architecture–social science connection seems to provide part of the basis for professional intervention. In extreme situations, architecture practice becomes a species of political mobilization work, public relations, or consent management and market research. All of this means new jobs (more producer services) and new specializations (new categories in the division of labor) on the design team.

Architects, social and behavioral scientists, and various combinations thereof function within today's construction economy as experts on the new people, the users and occupants of buildings. Not only has the successful construction of architects' new people led to a new category in the overall division of labor, but each of the design process stages represented in the ideal model has tended to become a separate compartment in the new subdivision of labor. Postoccupancy evaluation and the production of specialized social science knowledge and research for use in environmental design has become a respectable field in and of itself. Most building of any consequence involves extensive needs research, ranging from the home builder who does conventional market studies to define a salable "product," to the multinational corporation that examines the workspace "needs" of its clerical employees in order to help maintain a stable and productive labor force.

Programming has become a distinct occupation within design practice, particularly among larger public and corporate institutions. It is a distinct stage of work and has given rise to specialist firms that provide these services. These same corporate and government buyers of design services have in recent years expanded their middle-level professional staffs to include architects who function primarily as negotiators, managers, and representatives in the increasingly complex interpersonal activity that surrounds construction or development. Even the work of actually doing design in the formal or aesthetic sense, the turf most tenaciously defended by the traditional architects, often has its new microlevel division of labor, so that the design team includes talent trained or experienced in working with the social scientists, the programmers, and others among the many new participants. In fact, the time has arrived when the authorities accrediting architecture education and defining the content of licensing and registration exams include a small but significant amount of material on "social factors."

The possibility that the architects' new people vanguard played a key role in these economic transformations may be among the provoking insights that emerge from this recent history of architecture practice. Could the enthusiasts for user needs and an occupant-oriented design process have created jobs for themselves? This flies in the face of received economic doctrine, which holds that the workers did not create the industrial revolution, the capitalists did. Labor, in the traditional view, cannot create the market for its services, but does this hold true for professionals? The nineteenth-century anarchist, Mikhail Bakunin suggested that emerging large-scale industrial society would permit the rise of a "New Class" of bureaucrats and professionals who could translate their command of process and knowledge into power. We may now be seeing this power expressed in the labor market.[37] Following this line of thinking, it seems reasonable to suggest that the innovators and advocates responsible for articulating the architects' new people played an active role in creating a market for their own services. Is it possible that under the banner of the new paradigm the architects collaborating with social scientists have themselves set in motion the demand for more architecture inputs in a given quantity of built environment, thus generating some of the astonishing recent growth in architecture employment?[38]

"New" Architects' New People

The story briefly told in the previous section indicates that people work—that is, work that depends on managing interpersonal and intergroup relations—now plays an essential role in almost all phases of the construction and architecture processes. Seemingly endless ramifications of such work have become routine. These include not just the endless public hearings associated with the newly diffused levels of political interest and participation, but such evident changes as the new importance of marketing architectural services and the participation of the designers in a multisided interchange with financial participants. Architectural practices now often include marketing specialists and persons specially skilled in handling public meetings. Clients buy architectural services with an eye as much to their effectiveness in managing a conflict-ridden, litigious construction process as to the aesthetic or technological—or user needs satisfaction—dimensions of the final product. In this lies the splendid irony that concludes the story of the architects' people. The innovators were right all along that the architectural process had to make people the objective, but they were wrong about the identity of the people. *The architects' real new people were not the users and occupants, but all those who participated in and provided the context for the construction process.* The new paradigm erred in the same way the old paradigms of architecture as art or building science had erred. Both put the emphasis on the product rather than the process of construction. Major secular changes have forced a shift in emphasis. The economics, politics, and social context around construction have done this. Coordinately, and perhaps as much by default as by intention, architects have taken on major responsibilities in managing this context. As construction becomes more an economic, political, and social process, architects have found new roles, new jobs, new power as people workers.

Returning to the unanticipated employment effects of the social science–architecture conflation, in today's service society, the vigorous elaboration of the division of labor has become a central process. With the individual increasingly dependent upon work for personal identification and with higher rates of labor force participation, new jobs to which society ascribes worth become a key societal objective. It is an objective perhaps more important than more tons of steel, more cars, and more buildings. Work in and of itself takes on a value quite aside from its role in commodity production. A new and legitimate dimension of architecture concerned primarily with the new people, not with built form, carves out—or better, adds on—a new stratum of jobs in the construction economy. To confound things further the new people category is not limited to the users and occupants who were the limited objects of narrow, modern movement functionalism. Now the architects' new people, no longer merely objects but subjects as well, include all the people around the building process, especially the architects. This volume confirms this. Hidden in the talk about the architects' people lie clues no one can miss. What really excites these scholar-authors is the architect herself, as subject, in her role within the social process that produces the built environment.

This seems entirely consonant with the material facts of the rapid rise in interesting, highly regarded service jobs. Indeed, this trend stands out as one of the strengths of the late-twentieth-century American social economy, with its increasing deindustrialization and increasingly unequal distribution of wealth and income. Though some architects under the influence of boom times and postmodern enthusiasms may refuse to acknowledge it, the architects' people project has created a new subprofession and engrossing, well-regarded, socially useful work. And who knows, the ideal paradigm may lead to better buildings after all.

Notes

1. Charles A. Jencks, *The Language of Post-Modern Architecture* (New York: Rizzoli, 1977), p. 9.
2. The word *users* appears in quotation marks at this point to emphasize the rather special meaning the word has in current architectural talk. Those who developed the new ideas had in mind only a very specific and limited subset of all the building users. The architect certainly was not considered a user except in the special case that she designed her own office or dwelling. The fact that the architect used the building to earn a living did not count. The same applies to builders and construction workers, to owners who did not occupy ("use") their building, to investors, and to many others who from their own point of view might think of themselves as users. By the term *user* the people who developed the new idea meant only occupants, tenants, patients, inmates, workers and like people. Only rarely were space managers and maintainers included as users. When put in the combined form, the words *user needs* generally denoted only the needs of the same limited set of people.
3. Why new people? As the editors of this volume point out in their introduction, we can infer that architects in their imaginations, even in the earliest times, must have peopled their building designs with at least abstract representations of people. These usually tacit representations, whether manifest or latent, constitute the original or "old" architects' people. The new ones under examination in this essay represent something different—a construct from the recent confluence of social science and architecture—explicit, defined, researchable people, something quite the opposite from tacit.
4. More egregious than Jencks' factual errors is his ignorance of, or disregard for, the substantial research effort devoted to this sad project and the reasons for its abandonment

and subsequent demolition. Intentionally or in ignorance, Jencks gets wrong all the reasons for its demise. As the considerable body of scholarship indicates, Pruitt-Igoe was abandoned and demolished as a cost-saving measure in view of the local housing authority's fiscal crisis. Behind the fiscal crisis lay major changes in the metropolitan St. Louis housing economy. See Lee Rainwater, *Behind Ghetto Walls: Black Family Life in a Federal Slum* (Chicago: Aldine, 1970); Eugene Meehan, *Public Housing Policy: Convention Versus Reality* (New Brunswick, N.J.: Center for Urban Policy Research, 1975); the same author's *The Quality of Federal Policymaking: Programmed Failure in Public Housing* (Columbia: University of Missouri Press, 1979); Roger Montgomery, "Pruitt-Igoe: Policy Failure or Societal Symptom," in *The Metropolitan Midwest: Policy Problems and Prospects for Change*, ed. Barry Checkoway and Carl Patton (Urbana: University of Illinois Press, 1985), pp. 229–43; Roger Montgomery and Kate Bristol *The Pruitt-Igoe Myth* (forthcoming); and Kate Bristol, "The Pruitt-Igoe Myth," unpublished paper, Department of Architecture, University of California, Berkeley, 1987.

5. The label *environmental design*, as a broad umbrella term referring to the whole spectrum of professions concerned with the design of the built environment, came into use during the same period that saw the person–environment relations field evolve. Under the environmental design umbrella architecture is so numerically dominant that the two terms are nearly interchangeable. In some usages the term *environmental design* refers specifically to architecture design based on the new paradigm. Perhaps the best example of this usage occurs in the organization named Environmental Design Research Association (EDRA), the major professional organization for practitioners and academics in the person–environment relations field.

6. "Person–environment relations" as a name for the field competes with other phrases such as "environment and behavior," "architectural" or "environmental sociology," and "architectural" or "environmental psychology." Each phrase doubtless has somewhat different connotations, and certainly each has a constituency, but this is not the place to try to unravel all of that.

7. Henry-Russell Hitchcock and Philip Johnson, *The International Style: Architecture Since 1922* (New York: Norton, 1932), p. 36. The quoted passage refers to architects "like Hannes Meyer," but the idea was general according to the authors, though in most cases less extreme. They wrote further that the modern movement "conception, that building is science not art, developed as an exaggeration of the idea of functionalism."

8. J. M. Richards, *An Introduction to Modern Architecture* (Baltimore: Penguin, 1940), p. 37.

9. Reyner Banham, *Theory and Design in the First Machine Age* (London: The Architectural Press, 1960), p. 321.

10. Gerhard Fehl, "From Berlin building-block to the frankfurt terrace and back: a related effort to trace Ernst May's urban design historiography," *Planning Perspectives*, 2 (1987): 194–210.

11. This and the following quote from Larry L. Ligo, *The Concept of Function in Twentieth-Century Architectural Criticism* (Ann Arbor, Mich.: UMI Research Press, 1984), p. 5.

12. Walter G. Gropius, "Sociological Premises for the Minimum Dwelling of Urban Industrial Populations," *Scope of Total Architecture* (New York: Harper & Row, 1955), p. 98. This essay originally appeared as "Die sozioklogischen Grundlagen der Minimalwohnung," *Die Justiz*, vol. 5, no. 8 (1929), Verlag Dr. Walther Rothschild, Berlin-Grünewald.

13. Leon Festinger, Stanley Schackter, and Kurt Back, *Social Pressures in Informal Groups* (New York: Harper & Row, 1950).

14. The tenacity of this idea is evident today nearly forty years later, when designers still justify aspects of a design because they will "promote interaction." In all the years since Festinger the question, "Why interaction?" has largely gone unasked and unanswered. Perhaps it represents a version of the drunk searching for his house key under the street

lamp. Since "interaction" seems to have some empirically demonstrated dependent relationship to design, and design among those who subscribe to the new paradigm must be justified on the basis of its behavior modification properties, designers justify their work with this all-purpose rationale. For a curious and hyperbolic example of this see Christopher Alexander, "The City as a Mechanism for Sustaining Human Interaction," working paper, Center for Planning and Development Research, University of California, Berkeley, 1966.

15. Robert E. Park, Ernest W. Burgess, and Roderick D. Mckenzie, *The City* (Chicago: University of Chicago, 1925) introduced both human ecology and the zone hypothesis.

16. The National Resources Board, its predecessors and successors, under the leadership of President Roosevelt's architect uncle, Franklin Delano, brought the two disciplines together over planning and urban policy in the Board's Urbanism Committee, chaired by Chicago sociologist Louis Wirth. The principal policy publication of this group was *Our Cities: Their Role in the National Economy*, Report of the Urbanism Committee to the National Resources Board (Washington, D.C.: U.S. Government Printing Office, 1937).

17. Robert K. Merton, "The Social Psychology of Housing," in *Current Trends in Social Psychology*, ed. Wayne Dennis (Pittsburgh: University of Pittsburgh Press, 1948), pp. 163–88. Merton refers to A. Goldfield, *Substandard Housing as a Potential Factor in Juvenile Delinquency*, unpublished Ph.D. dissertation, New York University, 1937.

18. D. M. Wilner, R. P. Walkley, T. Pinkerton, and M. Taybeck, *The Housing Environment and Family Life: A Longitudinal Study of the Effects of Housing on Morbidity and Mental Health* (Baltimore: Johns Hopkins University Press, 1962).

19. During the 1950s all these scholars worked important veins in the person–environment relations field. In many cases the most important or the most broadly representative of their published reports came later; this explains the date for some of the works in the following list: Anthony F. C. Wallace, "Housing and Social Structure: A Preliminary Survey, with Particular Reference to Multi-Story, Low-Rent, Public Housing Projects," Philadelphia Housing Authority, 1952; Edward T. Hall, *The Silent Language* (Greenwich, Conn.: Premier Books, 1959); Leo Kuper et al., *Living in Towns* (London: Cresset Press, 1953); William H. Whyte, Jr., *The Organization Man* (Garden City, N.Y.: Doubleday, Anchor, 1956); Erving Goffman, *Presentation of Self in Everyday Life* (New York: The Free Press, 1959); Roger Barker, *Midwest and Its Children* (New York: Harper & Row, 1955); Robert Sommer, "Designed for Friendship," *Canadian Architect* (February 1961): 59–61, the first of a series of papers synthesized in Robert Sommer, *Personal Space* (Englewood Cliffs, N.J.: Prentice-Hall, 1969).

20. Edwin T. Hall, *The Hidden Dimension* (Garden City, N.Y: Doubleday, 1966); Roger Barker, *Ecological Psychology* (Palo Alto, Calif.: Stanford University Press, 1968); Sommer, *Personal Space*.

21. Sim Van der Ryn and Murray Silverstein, *Dorms at Berkeley: An Environmental Analysis* (Berkeley: University of California Press, 1967).

22. Neal Deasy, *Actions, Objectives and Concerns, Human Parameters for Architectural Design. A Planning Study at California State University, Los Angeles* (New York: Educational Facilities Laboratories, 1969). Louis Sauer and David Marshall, "An Architectural Survey of How Six Families Use Space in Their Existing Houses," *Environmental Design: Research and Practice*, Proceedings of the Third Annual Environmental Design Research Association Conference, ed. W. J. Mitchell (Los Angeles: University of California, 1972), Section 1309, pp. 1–10.

23. H. M. Proshansky, W. H. Ittleson, and L. G. Rivlin, eds., *Environmental Psychology* (New York: Holt, 1970); William Michelson, *Man and His Urban Environment* (Reading, Mass.: Addison-Wesley, 1970); Robert Gutman, ed., *People and Buildings* (New York: Basic Books, 1972).

24. That this American product of the 1960s could be thus represented as a deterministic model seems entirely appropriate; this was the era of such model building. At this his-

toric juncture, systems analysis provided the organizing concepts for many social science—related professional practice fields such as urban planning, public policy, and health care.

25. Philippe Boudon, *Lived-In Architecture; Le Corbusier's "Pessac" Revisited* (Cambridge: MIT Press, 1969).

26. Clare Cooper, *Some Implications of House and Site Plan at Easter Hill Village: A Case Study* (Berkeley: University of California, Berkeley, Institute for Urban and Regional Development, 1965). Later published in revised form as Clare Cooper, *Easter Hill Village* (New York: The Free Press, 1975).

27. Roslyn Lindhelm, Russell Ellis, Clare Cooper, and Jacqueline C. Vischer, *Final Report of Evaluation: Social and Behavioral Factors in Architectural Education*, Department of Architecture, University of California, Berkeley, June 13, 1975.

28. "St. Louis Is Revising Housing Complex," *The New York Times,* (March 19, 1972), p. 32; "Pruitt-Igoe Town Houses Proposed in New Plan," *St. Louis Post-Dispatch*, April 30, 1972, p. 28A; "City Discloses New Plan for Pruitt-Igoe," *St. Louis Post-Dispatch,* June 15, 1972, p. 3A.

29. Oscar Newman, *Defensible Space* (New York: Macmillan, 1972). If anything connected with the person—environment relations field can be said to have charisma, this is it. From its original publication and review in the Sunday *New York Times*, April 29, 1973, p. 16–17 to its publication in Britain the next year and subsequent almost literal adoption as cornerstone of public housing policy there, Newman's work came to stand for the whole disciplinary and professional subfield. The fact that in the United States the financial support for his work came largely from the federal Law Enforcement Assistance Administration and the National Institute of Justice (meaning mainly penology) should raise a cautionary flag. Despite a dropoff in official police support, Newman's version of the field remains alive and well, at least in Britain; see Alice Coleman, *Utopia on Trial: Vision and Reality in Planned Housing* (London: Hilary Shipman, 1985). Coleman, following Newman, makes the questionable assumptions that no intervening variables have much influence on the transaction between places and people, and that in the main, people passively receive what their environment has to offer rather than actively define, select, and otherwise intervene to determine the transaction.

30. John Zeisel, *Inquiry by Design: Tools for Environment Behavior Research* (Monterey, Cal.: Brooks/Cole, 1981).

31. John Lang, *Creating Architectural Theory: The Role of the Behavioral Sciences in Environmental Design* (New York: Van Nostrand, 1987).

32. This and the following section, in contrast to the preceding sections, do not draw heavily on historical sources and the review of standard texts. They draw instead primarily on the author's own research into changes in the American architecture labor market. Since most of this work remains unpublished, few citations can be given.

33. Producer services, a term used in labor economics, refers to service activities both with firms directly engaged in production, such as in-house engineering, architecture, or accounting, and with outside firms, such as those in engineering, law, and architecture, which provide services to production firms. For an early exploration of the subject and its definition, see Harry Greenfield, *Manpower and the Growth of Producer Services* (New York: Columbia University Press, 1966).

34. Any insights that may be in this paper draw heavily on a theoretical level form Dietrich Rueschemeyer, *Power and the Division of Labour* (Cambridge, England: Polity Press, 1986), especially his Chapter 6, "The Political Economy of Professionalization," pp. 104–40.

35. Roger Montgomery, "Rapid Recent Increase in Architectural Employment," working paper, Center for Environmental Design Research, University of California, Berkeley, December 1986.

36. Unpublished research by the author indicates that a significant decline in earning relative to other professions and the general price level occurred during the 1970s.

37. Insights in this paper regarding the power of professionals such as architects to create markets come originally from Alvin Gouldner, my main mentor in these matters. See especially his *The Future of Intellectuals and the Rise of the New class: A Frame of Reference, Theses, Conjectures, Arguments, and an Historical Perspective on the Role of Intellectuals and Intelligentsia in the International Class Contest of the Modern Era* (New York: Seabury, 1979) and his posthumous *Against Fragmentation: The Origins of Marxism and the Sociology of Intellectuals* (New York: Oxford University Press, 1985), which gets back to roots in the anarchist Mikhail Bakunin.

38. Parenthetically in closing, under the reign of the new paradigm in today's world, note the instructive contrast between the enormous professional environmental design labor inputs into Los Angeles' skid row project and its associated new and rehabilitated SRO housing as against the minimum professional interest and involvement in SRO housing depicted by Groth in his chapter of this volume. Times have changed.

Index

Delano, Franklin, 279n
Delight, 87, 88
Della Famiglia (Alberti), 34, 38, 41
Della Pittura (Alberti), 28, 42n
Delta Hotel, 218
De Menil Collection, 251–52
De Mille, Cecil B., 249
Democracy, Usonian, 14, 47, 187
De Officiis (Cicero), 25–26
De Re Aedificatoria (Alberti), 28
Derrida, Jacques, 166
Determinism, 225, 265–66
Dionysian-Apollonian dimension, 190
Diphilus, 26
Disjunctive-conjunctive dimension, 192
Dissimulation, 253–54, 257
Dr. Jekyll and Mr. Hyde (Stevenson), 197–98
Dog-trot house, 74–76
Domus, 23, 25–26
Doric columns, 23
Dorms at Berkeley (van der Ryn and Silverstein), 267
Dovey, K., 184
Downtown, 89, 90
    SRO tenants as citizens of, 215–21
Drama, Esherick on, 56, 57, 61, 63
Drucker, Johanna, 8, 161, 163–82

Easter Hill Village, 270
Eclecticism, 241, 243
Eco, Umberto, 62, 199, 250
Edison, Thomas, 225
Education
    Alberti on, 31–32
    Esherick on, 61–62
    Vitruvius on, 19–20
Eisenman, Peter, 9, 65–69, 90, 95, 97, 99, 126n, 161
    Drucker on, 163–65, 167–76, 180–82
Eliot, George, 79
Empire of Signs (Barthes), 155
Employee participation, 274
Encyclopedie, 244
"Enlightened" architecture, 185
Enlightenment, the, 196
Environmental competence, 190
Environmental Design Research Association (EDRA), 271, 278n
Environmental determinism, 225, 265–66
Environmental Psychology (Barker), 266
Environment and Behavior, 271
Esherick, Joseph, 14, 55–63, 161
Esherick, Wharton, 57, 58, 62

Existentialism, 184–85
Explorations in the Meaning of Architecture (Lennard), 198

Fano, Basillica at, 17, 21, 41n
Fascist architecture, 186
Favro, Diane, 9, 14–43, 211
Federal Housing Administration (FHA), 231–32
Fehl, Gerhard, 263
Festinger, Leon, 264–67
Feudal Society, 185
Financial sector, 273
Fire-water dimension, 190
Firmitas, 20, 23, 28
Flagg, Ernest, 229, 234
Flophouses, 221
Fontainebleau Hotel, 254
Foro Mussolini, 186
Forster, E. M., 57
Foucault, Michel, 67, 147, 166, 186
Fowles, John, 200
Frampton, Ken, 90
Frankenstein (Shelley), 197
French Lieutenant's Woman, The (Fowles), 200
Freudian psychoanalysis, 166
Frick Museum, 66
Function
    Beaux Arts theory of, 112
    monumentality and, 243
    transcendence of, 67
Functionalism, 118, 119, 211, 212
    humanistic, 126
    modernism and, 261–64, 272
    new paradigm and, 270
    program-driven, 125
    social science and, 264–67

Gainsborough Studios, 90
Gandelsonas, Mario, 167–68, 170, 171
Gans, Herbert, 257n
Gass, William, 65, 66
Gaudí, Antoni, 191
General Motors Technical Center, 127n
Generative grammar, 169
Genette, Gerard, 165
Gentleman's House, The (Kerr), 103, 131–36, 140
Georgetown Park shopping mall, 251
Gestalt psychology, 106
Getty Museum, 251
Giedion, Siegfried, 125, 243